SPINDRIFT

TRUE TALES FROM SCATTERED PARTS OF THE PLANET

by

Brian Hancock

PUBLISHED BY GREAT CIRCLE PRESS

SPINDRIFT

True Tales from Scattered Parts of the Planet

Written by Brian Hancock

Book layout by Brian Hancock

Cover design by Brian Hancock

Photographs © Brian Hancock

Published by

GREAT CIRCLE PRESS

9 Evans Road

Marblehead, MA 01945

United States

781-639 2904

greatcircle@mediaone.net

620 Mill Street

Newcastle, Ontario L1B 1C1

Canada

905-987 5735

Library of Congress Catalog Card Number: 99-96460

ISBN 1-929303-00-9:

US $16.95 softcover

CAN $24.95 softcover

Released simultaneously in the US and Canada

Manufactured in the United States

1 2 3 4 5 6 7 8 9 10

January 2000

For Tory whose questions inspired this book. And Sigrun and Brendon who have watched it take shape.

Also, for my father.

"I would rather be ashes than dust! I would rather that my spark should be burned out in a brilliant blaze than it should be stifled by dry rot.... the purpose of man is to live, not to exist. I shall not waste my days in trying to prolong them, I shall use my time."

Jack London

spin • drift   (n)   spray blown along the surface of the sea

spin • drift (story)   true tales from scattered parts of the planet

1- A Warrior Spirit

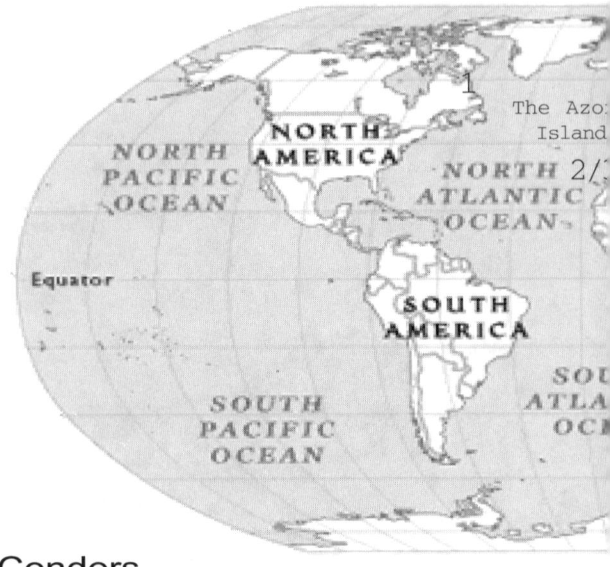

8 - Tea With Condors

9 - Beagle Channel Christmas

5 - A Serendipitous Piss

6 - Meeting Nando

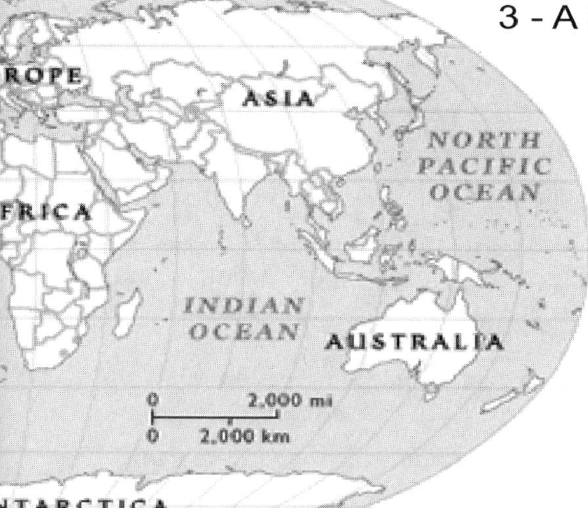

ARCTIC OCEAN
EUROPE
ASIA
NORTH PACIFIC OCEAN
AFRICA
INDIAN OCEAN
AUSTRALIA
2,000 mi
2,000 km
ANTARCTICA

ZIMBABWE
MOZAMBIQUE
Kafue R.
Zambezi R.
ZAMBIA
80 mi
80 km
Lake Kariba
Kariba
Karoi
10
Mount Darwin
ANGOLA
Victoria Falls
NAMIBIA
Chinhoyi
Mutoko
Sanyati R.
Harare
Inyangani 8,504 ft (2,592 m)
Chegutu
Hwange
Kadoma
Rusape
Mutare
Kwekwe
Chirhu
Gweru
Mvuma
Nsiza
Shurugwi
BOTSWANA
Bulawayo
Masvingo
Chipinge
Plumtree
Gwanda
West Nicholson
Chiredzi
Rutenga
Save R.
Beitbridge
MOZAMBIQUE
Limpopo R.
SOUTH AFRICA

SOUTH AFRICA
ZIMBABWE
BOTSWANA
Save R.
150 mi
150 km
Messina
Tropic of Capricorn
NAMIBIA
Pietersburg
LeBowakgomo
Olifants R.
MOZAMBIQUE
Limpopo R.
Johannesburg
Pretoria
Soweto
Germiston
Vereeniging
SWAZILAND
Piet Retief
Kuruman
Vaal R.
Upington
Kimberley
Welkom
Newcastle
Giant's Castle 10,876 ft (3,315 m)
Ulundi
Atlantic Ocean
Bloemfontein
Ladysmith
Alexander Bay
Orange R.
LESOTHO
Pietermaritzburg
Springbok
Prieska
Durban
Bitterfontein
De Aar
Middelburg
Vanrhynsdorp
Cradock
Umtata
Queenstown
Indian Ocean
Vredenburg
Worcester
Oudtshoorn
East London
Cape Town
Mosselbaai
Port Elizabeth

Throughout this book you will find mention of the Whitbread Round the World race. The Whitbread, as it's commonly known, is a thirty-thousand mile circumnavigation of the globe. From humble beginnings in 1973, it has grown to become the preeminent ocean race, attracting top sailors from around the world and pitting them in fierce competition that sends the fleet through torrid heat, and blinding cold. It has become a proving ground for sailors, and is a must-do event if you plan to rise to the top of the sport.

I raced my first Whitbread aboard *Alaska Eagle*, a 65 foot American entry, and returned again four years later to sail aboard *Drum*, an 80 foot British entry. In 1989 I sailed with the soviets aboard their entry *Fazisi*. All three trips were interesting, challenging and very different.

The Whitbread is a tough way to make a living, but the rewards are huge. The pure sense of accomplishment at finishing the race rapidly outweighs the trials and tribulations of making the trip. The cold, wet, windy days soon become distant memories, the misery quickly forgotten, replaced by a sense of achievement which stands foremost in your mind.

***

You will also find some references to "we" and "us" when I am sailing alone on the boat. This might be confusing because you will wonder who else is on board. I am alone, however I always refer to it in the plural because the boat has as much to do with the sailing as I do, and it only seems right to include it. We are partners – often I hear it talking to me and when it does, I listen.

FOREWORD by Skip Novak

In the summer of 1979 I was due to skipper the sixty-five foot ketch *Independent Endeavour* on a commemorative race from Plymouth, England, to Fremantle, Western Australia, with a stop in Cape Town.   I had found most of the crew by July, most of whom I had sailed with before. We were all in our twenties and about to embark on a great adventure.   We were still lacking a sailmaker for the trip, and I was trawling the waterfront for candidates.

"Anybody around?" I asked a trusted Kiwi skipper who had just crossed the Atlantic.

"Yeah, try Brian Hancock, he's a South African, still wet behind the ears, a smartass, but a good shipmate. We picked him up in the Caribbean, just up from Cape Town. That's him on the dock there repairing a sail cover." I walked over.

"Interested in sailing to Australia via South Africa in September?" I asked, as he cranked away on a manual sewing machine. He looked like a kid.

"Sure, what's the deal?" he answered with a confidence that belied his age.

"It's the Parmelia Race. As usual there is no deal. You can join us next week."

"Why the hell not. I'll be there," was his quick reply.

Brian Hancock, or Mugsy as he was know then, was twenty-one years old. He had just signed on for a three month voyage after a five minute conversation. No pay or other prospects were forthcoming, but that was the way it was done back then in what I call the 'Golden Age' of ocean racing. Things were generally casual, sometimes haphazard, at times reckless. In short, it was all about youth.

So began a twenty-year friendship that among other things, has put over seventy thousand miles of blue water under a common keel. We sailed together on *Alaska Eagle* in the Whitbread Race in 1981/82, through the Southern Ocean on *Drum* in 1985/86, and to Uruguay with the Soviets on *Fazisi* in 1989. There were many other minor voyages as well.

We hit if off from the start, both ironic, able to laugh at people's misfortunes and appreciated the differences in race and nationality with a benign sense of humor which today is considered very 'politically incorrect.' We both came from disenfranchised backgrounds. If you have travelling in your blood from the beginning, Chicago, where I grew up, is just as restrictive as Pietermaritzburg, where Brian grew up. When other twenty-year olds were contemplating careers as doctors or lawyers, or planning business empires, people like us, and god knows there were many, had one simple agenda – to see the world. In both our cases, sailing was our ticket out. Joseph Conrad was the author of choice.

Those were the days when you lived from a seabag and could move at twenty-four hours notice – to anywhere on the planet. We were nomads. Those in the genre blew with a metaphorical wind taking the path of least resistance, while paradoxically searching for another epic challenge. We were opportunistic, certainly not for material gain as there was little, and also not for reasons of climbing any ladder – no, we were instead hungry for the next bold move, the next lot of shipmates to laugh at or with, always anticipating riding that next wave.

For an ocean sailor, it is all about that moment of toughing it out in heavy weather with all hell breaking loose and knowing that you're still in control. At other times it's more serene, simply being at the wheel while carving a long surf in the deep blue. Brian was one of the best I sailed with, having both an appreciation for the serenity and a competence for the storms. The many sensations you experience at sea focus your mind in order to rest your soul – this in turn gives you the energy to want to do it all over again. It's an addiction. It's risk. It's a simplicity. We can't imagine a life without it.

Robert Service, in "The Men That Don't Fit" wrote, "it's the steady, quiet, plodding ones who win the lifelong race." I'm sure he's right. Here we are in middle-age, landed, but not grounded. Brian is still scheming and planning the next move. In fact he called me this morning to discuss plans he has for skippering a big catamaran in the Cape to Rio race. I am the same. Long term commitment in the workplace is an anathema, freedom everything, or at least a sense of it. Our seabags have of course grown. We now have houses to keep them in. E-mail is the secret to our subsistence. Somehow, with no foundation whatsoever, we just know that things will turn out alright.

Common experience in the field, that's the thing. The travelling by land or by sea, is the thread that has kept us together. Frankly, I think I would rapidly tire of Brian in a 'drawing room' relationship, and he of me. So we keep voyaging, both tied to our separate helms, reaching hard with a driving spray in our face. We wouldn't have it any other way.

I hope you enjoy his book – it's a good read.

TRUE TALES (a preface of sorts)

*"Listen attentively, and above all remember
that true tales are meant to be transmitted;
to keep them to oneself is to betray them."*

Baal-Schem-Tov

Writing a book is a self-indulgent exercise. Expecting someone else to take the time to read it requires a lot of nerve. Asking that person to pay good money for the privilege surely places writers among the most ego-driven individuals on earth, yet there are millions of books written each year, and most of them are read.

I hope that you will read this one.

I hope that you will take a moment, leave your day behind, and join me on a trip around the world.

We will sail across the Atlantic, hike through Africa, climb mountains in South America, round Cape Horn under sail in a full gale, and spend some quiet time with the locals. It will be fun. Don't bring anything – I have it all taken care of. We will eat out in Brazil, enjoy a glass of South African wine, sip tea in Tierra del Fuego and try a surprise drink south of the southern-most town in the world. If you like adventure you will like this book. If you don't, you will still like this book. It's about life and living it to its fullest. It's about human frailty, and its ability for recovery. It's about people and friends and family, and how we all make this a great place to be from – planet earth that is.

So join me. These are true stories with a bit of poetic license tossed in. They were started for my daughter who grew up with a father who was always away someplace on some adventure. I wanted to leave her a record of what I had done, hoping perhaps that she might be inspired to have adventures of her own. They have evolved into a book. The stories are not in chronological order. "A Date With a Wall" is a sequel to "Remembering Eric," and "Meeting Nando" continues "A Serendipitous Piss." The book starts off and finishes in Newfoundland where I traveled to gain some perspective on my life after struggling with a mild depression resulting from an incident related in "A Date With a Wall." Other than that, each story stands alone. "Spindrift" and "Same Thing Different" are short essays. They are my take on life, especially life at sea.

Take your time reading this book. Pass it around, and when you are done, head off on an adventure of your own. Live a full fun life. But be warned: there is no going back. Once you have tasted the open road and seen the fresh dew on a new day, you will be hooked for life, and there is no antidote.

There is a Chinese proverb that states: *"A voyage of a thousand miles starts with the first step."*

Take that step and thanks for reading.

A WARRIOR SPIRIT

"Mother, mother ocean, I have heard your call
I've wanted to sail upon your waters
since I was three feet tall
You've seen it all, you've seen it all."

Jimmy Buffett

The cold waters of Newfoundland are a good place to think, and in the summer of 1999 I needed time to think. My otherwise charmed life had found a fork in the road, and for the first time in forty years I was following a path through uncharted territory. I was traveling down the dark side of my mind. The side where doubts feed upon themselves and grow in your imagination. The side where confidence can be crushed by a single word. It was not a good place to be, but a mood of melancholy had settled on my soul and I couldn't shake the feeling. I needed to get away. I needed to spend time where the world was unspoiled and allow my thoughts to flow freely. I needed to purge my memory of a recent incident that had unsettled my settled life, and replace the space with new memories. So I traveled to Newfoundland to go sailing.

I saw the first iceberg from my seat on the plane as we flew towards St. Anthony on the northern peninsula of the Canadian province. It was bright white, the billions of compacted snow crystals reflecting the afternoon sunshine. As I watched the berg disappear under the wing of the plane, I knew that I was heading in the right direction, and as we touched down on the small airstrip I was sure I had done the right thing by getting away. The air was cool and clean. I breathed deeply and felt tension leak from my body. By the time I stepped on board the boat an hour later, the familiar surroundings of a well appointed yacht were a little like coming home. The warm teak trim and comforting slap slap of water on the hull were old friends. I cherished the feeling and knew my luck was changing. I had recently returned from sailing across the Atlantic in an effort to dispel a streak of bad luck that had been plaguing me, and as I looked around the small Newfoundland fishing village, I knew I had kicked the curse. At least I felt I had, and to me that was as good as the real thing.

We would sail the rugged northeast coast, slowly making our way towards the capital city of St. John's on the east coast, and then passage across the Grand Banks to Nova Scotia. It was a short hop from Halifax to the boat's home base of Dark Harbor, Maine. As night fell on my first night away, I heard the clear mournful lilt of a loon. It sounded like I felt. Sad to be away from my family, (I was recently and happily married) but happy to be on an unspoiled acre of the earth. The charm in my life was slowly returning. We would leave in the morning.

***

I am not the type who views age as a measurement of one's life, and so I surprised myself when I faced turning forty with some trepidation. It was not like me to be looking in the mirror and asking myself hard questions, but there was a restlessness that lurked in my gut and I couldn't place it. I had lived a rich and varied four decades and felt fortunate to have traveled the world. I had seen many diverse people and enjoyed their cultures. I had sailed southern seas and hiked high mountains. I had seen a lion pad silently past my tent in the night. And then one morning I woke up and it all became clear – I was bored. I needed another adventure; something big. Something by which I could measure everything else in my life. In my early twenties I sailed my first Whitbread race and it had given me a yardstick by which I could measure the days and months that followed. The time turned into years, but despite repeating the trip a few years later, the memories slowly faded and with them went my measuring stick. It was time to do it again, but this time it had to be a new adventure. I settled upon sailing around the world one last time, only this time I would do it alone. It wouldn't be easy. It would require all my effort and creative resources. I would definitely need some help from lady-luck, but I knew I could count on her. She always rose to the occasion. The only thing that stood between me and my goal was a million dollars. My immediate problem was that I could sum up my financial situation with a single word; I was broke. Then lady-luck intervened and a month later I was the proud owner of a fifty-foot carbon-fiber racing yacht built specifically for solo ocean racing. Once again my charmed life was living up to its reputation. It had happened so often that I was starting to take it for granted; something I would soon come to regret.

***

We left St. Anthony and sailed smack into iceberg-alley. On the horizon ahead a massive tabular berg was drifting slowly

We sailed towards an angular chunk that lay directly in our path to our evening anchorage.

south. There were smaller bergs floating closer and we set sail for an angular chunk that lay directly in our path to our evening anchorage. The bergs had been spawned from the arctic icepack three years earlier. The glaciers had calved virgin bergs into

Greenland Sea where they were picked up by the Labrador current and swept south. The ice sails in the cold water, traverses from the Greenland shore to the Canadian coast, and converges at the Straits of Belle Isle. When they pass St. Anthony, their steep sides show deep striations where warmer weather and relentless seas have taken their toll. They balance precariously on an eroded underside and capsize with regularity. We approached our berg like typical tourists, and launched the dingy for a closer inspection. We were just approaching the ice when the side sheered off and crashed into the sea. The berg listed heavily, and for a moment it rocked gently, threatening to capsize, but just as quickly it settled back on its axis as we scurried for the relative security of the boat. We were learning the ways of temperamental ice and gaining respect for its timeless beauty. My accident off Bermuda seemed a far-off memory, and when a bald eagle buzzed us, I knew that I was on the mend.

When I do things, I do them from my heart. My head rarely enters into the equation. I need to feel the idea has a strong footing deep in my gut, and when I feel the familiar lump, there is no going back. It has served me well for most of my life. It helps that I am a dreamer and an unflaggable optimist because my path is littered with grand schemes that any reasonable person would have deemed dumb, and most reasonable people deemed sailing around the world alone dumb, but I was not deterred. The critics had said the same about my first two circumnavigations. I thought it was the grandest idea I had ever concocted and set about preparing for my trip. I would be an entrant in the Around Alone Race, a quadrennial event that pits sailors from around the world against each other, and, more importantly, themselves. It is called the toughest race in the world and it probably is. In order to qualify you have to sail across an ocean alone, on a passage not less than two thousand miles, and it was at the end of my qualifying passage that it all fell apart.

***

My trip to Newfoundland definitely found its roots in my heart. I have always been attracted to remote places and the steep-sided fjords and rugged coastline has long held a fascination. So too have the people that inhabit the island. I have tremendous respect for the qualities that bind people together, especially those that live in an uncompromising environment, and the spirit and generosity of the people we met along the way were a true testament to their nature. A week into the trip we found ourselves wedged up a fjord in an anchorage off a small fishing village. The water was still and the colorful cottages reflected gaily in the fading light. As the last rays of day seeped from the sky we heard clear voices singing out to us. Two young girls from the village had walked to the waterfront, and were singing traditional Newfoundland songs. Their sweet voices carried across the calm water and amplified on the steep sides of the fjord. They sang until it was dark while we sat and listened to songs of love and life drift gently on the evening air. They sang of their people, of loves lost, and they sang of fishing and what might have been. They sang of sailing, the lyrics painting vivid pictures of schooners wet with spray and heaving seas pushing them towards home. They sang about things that are important in my life, and each song eroded past memories and replaced them with fond new ones.

The next day the wind was honking from the south and we sailed close-hauled along the coast. Arctic terns and

Atlantic puffins kept us company. We laughed at the frantic flight of the puffins – their frenzied flapping barely keeping them airborne. Numerous humpback whales blew fishy air our way, while they cruised slowly by, each following an ancient calling. It was cold, and layers of warm fiber-pile kept the chill from my bones. For the first time in nearly a year I felt as if my life was back on track. The fingers of self-doubt that had flirted with my mind over the winter were washed away by the cold air, and I was content again. Perhaps I had set myself up for failure by attempting the impossible. Perhaps I should have hedged my bets in case of failure. It was no use second guessing myself. I had dealt from my heart and felt the sting of failure in the same place. It stung harder because it was the first time. It stung harder because I was not ready for it. I sat and watched the spray from the bow wave wash the foredeck and thought about all the grand schemes I had attempted, and how many of them had worked. My ability to seize an opportunity and turn it into a reality had served me well. My life was full of stories of travel and adventure, and I reflected for a while on all the funny and dangerous incidents that had taken place. I had them all written down. Those that were not committed to paper were stored in my memory bank. Not all of them had been successes, in fact the really good stories were about follies. Collectively they were what got me to where I was today and it had been an interesting life. It was time to revisit the stories to see what lessons they held.

As we anchored that night I felt the last of the demons fly from me.

We rowed ashore to a small beach littered with driftwood and made a bonfire. The flames licked at the wood and the glow shone on the faces of my crewmates. I had returned to the simple things in life. The things that give me strength.

A full moon rose behind a forest of birch trees casting its bright light on the still water. As I watched its slow transit of the sky I remembered a poem that I had once been given. The words rang true. Three short sentences that sum me up.

"Some say the young boy became a sailor
that his warrior spirit carried him to distant shores…

I remember seeing him once on the southern shore of Africa
His eyes caressing the canvas…

I thought I heard him whisper his dreams to the southwest wind."

I have always whispered my dreams to the wind, and they have always come true. They did this time along the rocky shores of Newfoundland.

REMEMBERING ERIC

"Have you gazed out on the ocean, seen the breaching of a whale
Have you watched the dolphins frolic in the foam?
Have you heard the song a humpback hears five hundred miles away
Telling tales of ancient history – of passages and home?"

John Denver

I'm at sea again. It's the summer before my trip to New-foundland and I'm aboard my own boat four hundred miles east of the Azores. The sun just dropped into the ocean behind us spluttering and sizzling as it set, leaving a starlit night and a lumpy sea ahead. I pulled the fishing line in and we settled on canned spaghetti for dinner. Other than the gale that blew through a few days ago, this has been an easy passage, marked by a series of strong lows feeding us a steady westerly flow. My boat is beautiful and rides the wind like a cork full of jet fuel. It is designed for this kind of sailing and flies along oblivious to occluding isobars and impending forecasts. It's a night to remember, and as I head below to write up the log I glance at the instruments mounted above the companionway. Boat speed is ten knots. I feel a fine spray from the bow wave and duck as solid water splashes the coach roof. The red night-lights illuminate the interior. I settle into my nav seat, flick the single-sideband radio on, and tune in to the BBC world service.

"This is London," the rehearsed voice announces, and the theme music transports me back thirty years. For a few moments I am a small child at dinner, eating in silence while my parents listen to the world news on the BBC. It was a mealtime ritual in our house, and stories of war-torn villages and drought-ridden areas went hand-in-hand with lamb chops, peas and mashed potatoes. Five sharp beeps, one long tone, and a voice announces.

"The search for a well-known yachtsman presumed missing off the coast of Wales, has been called off until morning. Authorities announced this evening that they had suspended their search until daybreak tomorrow for the sailor who apparently fell overboard from his yacht. In other news…"

"In other news," I say out loud. "What other news?" Someone I know, or know of, is missing and I wish I knew who they were talking about. I would have to wait until later in the broadcast to find out who it was. While I wait, I tip my chair back and gaze out the window. My nav seat is comfortable and the view from inside the cabin is outrageous. I have a clear panorama of the world to leeward, and watch the swells drift lazily by. A sliver of moon reflects on the water casting dancing shadows on the waves. For a moment my thoughts are back on the news, and then in a flash I see dolphins. Actually I hear them before I see them. There is a sharp squeak and a rattle, followed by quick exhale and a blur as two Bottlenose dolphins fly by my window. They hit the water without a splash and are gone.

My crew are in the cockpit. They are in deep conversation and have seen nothing.

"Dolphins," I say as I clamber over them and head for the bow.

Great Circle — my cork on jet fuel.

"Come on guys, let's see you jump." I yell into the night. The ocean below me is still and I watch the bow rise and fall with each swell. It knifes through the water and cuts a clean path across the surface. I hang onto the bowsprit and watch the waves for movement. "Come on guys," I plead. This passage has been void of wildlife and I am aching to see another sign that life exists. I hang over the bow until I can just see the keel, and they come at me like children playing with a water hose. I hear the squeaks and watch a dozen dolphins fly by on the lip of the bow wave. The water in front of us is alive with wet flying mammals, and they slice the surface and disappear into the clear dark water. My boat responds in kind and surfs down the edge of a short Atlantic roller. And then, just as soon as they came, they are gone. I stare at the swells and hope that they will reappear, but without luck. Another moment of instant magic vanishes into the vast open ocean.

I duck below and search for better reception on the radio. The broadcast is scratchy, but it's the way I like it. It reminds me that I am at sea where the news happens to other people. We are in our own world out here, and the feasts and famines from other parts of the planet have little effect on us. Marine forecasts and squalls on the radar are our news. The commentator announces an impending drought in Somalia, and then continues;

"The famous French sailor, Mr. Eric Tabarly has apparently been lost at sea off the coast of Wales. Mr. Tabarly was well known for his successes in single-handed yacht races in the 1970s, and was a veteran of a number of Whitbread Round the World races."

I stare at the radio in disbelief. Tabarly, that great icon of French sailors, is gone. Lost at sea. Three short words that no sailor ever wants written after their name. I can't believe what I have just heard. The commentator continues:

"According the local authorities, Mr. Tabarly was sailing aboard his yacht, the Pen Duick, and fell into the water while

changing sails. He was not seen again. Attempts to get help were hampered by a faulty radio, and it was only when an Australian yacht passing by was hailed, that the news was passed on to authorities. Search operations have been called off for the night, but will resume in the morning. Local officials are calling the drowning accidental, and are certain that he will not be able to survive the night in freezing temperatures. They hope to recover the body in daylight. " I stare at the radio for a moment longer and then flick it off. I am not interested in the cricket scores.

"Damn," I say to myself, "Tabarly's been lost at sea. Gone forever."

The thought sobers me.

Eric Tabarly paved the way for single-handed sailors, and was in a small way responsible for the passage I was now making. I first met him during the 1981 Whitbread race, and saw him many more times after that. Where there was a big sailing event, Tabarly was there with his characteristic Gauloise dangling from his mouth, and twinkling eyes taking in the scene. He inspired legions of young French sailors, many of whom rose to the top of the game to win all of the single-handed races around the globe. Those sailors, in turn, inspired me to join the world of solo sailing, and here I was en route to the Azores aboard my own single-handed sailboat. For the ride over I have crew. I will return to the US alone. Eric, if you are up there, keep an eye on me will you? You're partly responsible for this lunacy.

I stare at the radio for a long time. From my nav seat I can see the wind instruments at the top of the mast. A small light illuminates the wand and I can see that the wind is just aft of the beam. The motion of the boat is gentle as it rises and falls on each wave. There is a continual creaking from the halyards, and an audible moan from the autopilot. They are comforting sounds. My instrument panel is like that of an airplane and I can monitor just about everything from my seat. I glance at the radar and it sweeps a clean scan. The GPS* blinks a speed over the ground of nine knots, and puts our ETA in Flores in forty-four hours. I note the details in the log and check the screen one more time. Speed is up and our ETA jumps forward five hours. We are still way out in the Atlantic with a long way to go. A lot can happen so I disregard our arrival time. I tilt my seat back and shut my eyes. The news of Tabarly's death stings. One of sailing's best has been lost at sea. If it can happen to him, it can happen to just about any of us, but then I knew that already. I squeeze my eyes together and try to remember his face and the last time I saw him. It was in Uruguay five years ago. Racing another Whitbread.

***

There is one story that sums the man up, at least for me. I am sure that there will be many more told at his memorial service in Paris next week, but I will remember him most from an encounter in England the day before the start of the 1989 Whitbread. I was racing that year with the Russians aboard *Fazisi*. The boat was built rough. She had a low slung freeboard and no headroom. There was little thought given to comfort, and the bare aluminum showed scars where the panels had been welded. The head was tucked under the companionway and its users would require lessons from a contortionist. I had arrived in England the day before the start and was making an attempt to stow my gear and get familiar with the layout. There would be

*GPS - Global Positioning System, a navigation system that uses satellites to pinpoint your position anywhere on the earth.

Skip Novak and myself, and fourteen Russians for the trip to Uruguay. The crew were bolting down bits of deck hardware, and welding torches and electric drills were being pressed into overtime. I sat amongst the boxes of food wondering what the hell I had let myself in for.

A moment later there was a quiet tap on the hull and a voice with a thick Dutch accent called out. I went on deck to be greeted by Connie Van Reitschoten, the venerable Dutch sailor and winner of two Whitbread races. Connie had upped the ante when he came onto the Whitbread scene, and poured money and organization into his campaigns. They became a benchmark for future winners.

"Can I please come on board for a look around?" he asked. I held out my hand and he jumped the lifeline with ease. I scrambled down the aft companionway and waited in the nav area. Connie slid down the stairs, ducked to fit below, and shook his head in disbelief. His yachts had been finished by expert craftsmen, and a liberal use of teak added to the look of quality. He stared at the dull gray interior, not uttering a word. Clearly life on this Russian yacht was not for him. He stared for a long while, shaking his head, and then left. I watched him walk down the quay. He stopped to talk to someone, and I saw him point towards *Fazisi*. He was still shaking his head. I smiled to myself and returned below to finish unpacking.

A few minutes later there was another knock on the hull, and a French voice called out for permission to board. I went on deck to see Tabarly clambering over the lifelines. He shook my hand, tossed his Gauloise in the water, and asked my permission to go below. I followed him and watched his face light up when he saw the interior.

"Oui, tres simple, c'est bon," was all he said, but coming

from Tabarly it was tantamount to a blessing. He beamed at me, clearly impressed by the basic layout. He was a man whose spirit truly flourished when surrounded by simplicity. A man true to himself and his profession. The sailing community and the world

En route to the Azores — below at
the nav station writing this story.

in general will be worse off without his twinkling grin and sparkling eyes.

My own tendency falls on the side of Tabarly. I prefer simple and functional over ornate and impractical. I have sailed around the world on "floating furniture" and enjoyed the trip only when tied stern-to in a marina. My own yacht, *Great Circle,* is built for speed and solo sailing, and as a bonus she rides the ocean with ease and deliberation. This night on the open sea two thousand miles from Marblehead, my home away from home is flying through the dark night, silent as a ghost. The only trace she leaves is a sparkle of florescence that bubbles in our wake. A trickle of water on the hull hums as we accelerate, and the autopilot moans in response. If all goes well we will make Flores, the westernmost island in the Azores group, in a passage of less than ten days.

I can almost smell the flowers.

***

My crew jumped ship in Flores. The lure of life on land and the pressing need to attend to daily business, quickly snuffed any thoughts of continuing on to Horta. They took a small commuter jet to Punta Delgado, and in moments were back in the traffic and jams of the real world. I was left to my own thoughts tied to the quay in Flores while a front passed overhead. The wind whipped through the rigging and volcanic grit from the dock splattered the hull while I holed up in my bunk reading, waiting for a break in the weather. Every hour or so I climbed ashore and clawed my way to the edge of the breakwater to gaze at the waves crashing into the concrete pylons. I would rather have been riding out the storm far out to sea, but once tied to the quay I was reluctant to leave. Some yachts simply do better away from land, and *Great Circle* is definitely one of those yachts. My gut winds itself into a knot as soon as we approach solid ground.

By early evening the front was showing signs of easing and a short walk beyond the breakwater to a high point on the island confirmed that the sky was brightening to the west. I did not want to spend another night listening to the dock-lines tugging at the cleats, so I readied ship for departure. It took only a few minutes to secure below and crank up my outboard, and I left Flores as dusk descended on the island.

I was anxious to be away.

The sea was still lumpy and confused, but the wind had definitely dropped. Black rain squalls were lashing the higher elevations. To the west puffy pink clouds reflected the last light of day. The storm was over. Feeling a bit seasick and lonely I hoisted the main to the second reef and dialed in a course for Horta, 132 miles away. "I'll be there before dark tomorrow," I thought, but I had forgotten that islands create their own weather. I was going to pay for my impatience before this trip was over.

I spent a restless night scouting the horizon for ships and feeling the boat wallow with too little sail up. I did not have the energy to hoist the main all the way, and we flopped around with two reefs and a jib until a cloudy dawn rousted me from my lethargy. I set a full main and washed the decks to get rid of the grit. A large, even swell was running in from the southwest bumping up against the new wind from the east, and the slop and chop left me feeling queasy. With the end of the passage in sight, I was thinking about beer and hot food in Peter's Café Sport in Horta.

I could almost smell the garlic.

The afternoon wore lazily on as the wind came and went. I ran the outboard a few times. It was my only engine, but made no difference against the slop and I resigned myself to missing closing time at Café Sport. The breeze, or what there was of it, was right on the nose. I fell asleep while my carbon-fiber home bounced around until I could just make out Horta on the radar.

It was shrouded in fog and not visible from the deck. The air was still and moist. The radar hummed while I watched the blip of land remain steady at ten miles. Suddenly, on the opposite end of the screen, another blip appeared as the scan found a

The afternoon wore lazily on
as the wind came and went.

disturbance on the water. A squall I thought, and went up on deck to investigate. The western horizon was black with low storm clouds marching in formation. It was a solid wall. The radar screen confirmed my visual, but I was not concerned. We were close to getting in and I was ready for some rest. We plodded along with the autopilot driving, barely making three knots. I grabbed my foul weather gear and made a cup of coffee. The line marched closer; suddenly it was upon us. I felt the first fingers of wind whip the jib and the boat heel gently in response. There was a light rain, and I was pulling up my bib when the front hit. It slammed us with all the might of a moisture laden army and in a second we were rail down and spinning out. The autopilot alarm screeched a warning as I hit the standby button.

"This is all I need now," I thought. We were so close to land – then I looked at the radar again. The screen was a solid mass of interference. Whatever had been behind was now directly overhead and here to stay. The autopilot alarm screamed with more intensity and refused to relinquish the wheel. I jabbed at the buttons while water broke over the side of the boat, but the pilot held fast and I was unable to alter course. I jumped down the companionway and scrambled aft to the switchboard to hit the breaker. I could feel the yacht lurching and moaning under the load. I snapped the switch off and the alarm stopped. As I headed back I grabbed my harness and shot on deck grabbing the helm. It was still frozen. The pilot was off and the alarm was silent, but the helm remained jammed. "Oh Christ," I thought, "now what?" We were knocked flat on our side with the mainsail sheeted in and the jib still pulling. I hit the jammer* on the jib halyard and heard it let go. The sail flapped, but the force of the wind prevented it from dropping to the deck. I would have to drag it down. The deck was at eighty degrees and I inched my way forward bracing against the side of the cabin. I grabbed handfuls of sail and the jib came down quite easily. I

* **Jammer** - a cleat that holds lines under pressure. The line passes through the jammer and a series of clutches holds it in place.

lashed it with a sail-tie and inched my way aft to the cockpit again.

I still had to deal with the mainsail.

I wrapped the mainsheet around a winch before breaking its jammer open. The pressure eased on the main and the boat came upright. It wallowed for a second with the sail flogging, and then the autopilot had a change of heart. Its small computerized brain decided that it wanted a different course, and I watched the wheel slowly turn to port. The first fingers of fear tickled the base of my spine.

"Oh no, please don't do that," I said out loud. I watched the wheel turn in slow motion, and felt the boat lumber in response. The bow dropped away and the boat came upright. We took off on a reach. I hung onto the binnacle trying to stop the wheel from turning, but without luck. I watched as we bore off and the world started to spin a little slower. The wind died down, and the noise of the knockdown was replaced by a more gentle whistling in the rigging and the scream of water flying past the hull. I watched in horror as the boat took off down a wave and the speedo topped sixteen knots.

"Oh fuck, oh please no!" I yelled into the wind and then flung myself onto the cockpit sole grabbing the binnacle. I watched the speed climb to eighteen knots while the wheel continued its slow turn to port, and then with all the force of a five-ton yacht on the lip of wave, we gybed – all standing.

White water screamed past the boat, the boom flicked into the air, and then it swept the deck. It whistled over my head as I crouched wrapped around the binnacle, and with a huge crash slammed into the leeward running-backstay*. There was no way

* **Running backstay** - a series of lines that hold the mast in place. They are found on each side of the boat and used accordingly.

the rig could handle the impact. Either the mast, the boom or the backstay would have to break, and if the backstay broke, the rig would soon follow. In a split second we were back over on our other side and rounded up into the wind. I saw that the mast

The western horizon was black with storm clouds marching in formation.

was still in one piece just as it dipped below the surface of the water. We were over, right over, and the autopilot still had the wheel in a tight grip. I unclipped my harness and climbed below. The nav station was hanging loosely at an awkward angle, and the spaghetti from the stove was plastered on the bulkhead. I dragged myself through the narrow alley that separates the main cabin from the back of the boat, and found the steering quadrant. It was dark and wet and almost upside down, but I knew where the autopilot ram was connected, and searched for the nut that held it to the carbon quadrant. It was loose, and in a second it was off. I hit the ram and it came away easily. The wheel was now free from the computer brain and I scrambled on deck to figure out what to do next.

The rig was still slapping the surface of the water, but at least it was in one piece. I had to get the new running backstay secured before easing the old one. Without a secure backstay, the rig was as good as gone. I wrapped the tail around a winch and snugged it tight. The noise of the wind was ripping through the rigging and waves crashed over the boat. I looked up at the sky hoping Tabarly was up there watching me.

"Come on Eric," I shouted. "You've been here, give me a break," but my words were lost in the wind. Tabarly was still a newcomer up there and probably did not carry any clout.

I scrambled down to the low side to release the jammer holding the leeward runner, but the outer cover was ripped and was bulking up against the jammer feed. I could neither tighten nor release it until the cover was cut loose. With the mainsail plastered up against the running-backstay, we would remain pinned until I could do something about the cover. I crawled below again and found my large kitchen knife from home laying in the bilge. I grabbed it, and on the way back stopped by the nav station and opened my laptop to check my course. The electronic chart was blaring a warning. Our projected course was heading directly for land, five miles to leeward. A lee shore in a full gale. We were drifting at two knots.

I felt surprisingly calm. All I needed to do was cut the cover on the runner, ease the main, drop the sail, set a storm jib and alter course. It was simple to see, but looked impossible. I wrapped myself around a winch that was half under water, and slowly cut at the cover. Inch by inch it came free. Bits of spectra yarn littered the cockpit. They were soon washed away by the rain and waves. The knife was big and sharp, but I wished I had a pair of scissors on board. It would have been easier with scissors. Raindrops pelted the water and the mast continued to tap dance with the waves. It was a precarious situation, but surprisingly under control.

I heard the wave an instant before it hit us. It came with a low growl and slammed into the side of the boat tossing us over like a tiny cork. I saw the mast disappear below the waves again, and grabbed the binnacle. I felt a sharp sting on my thumb, but waited while the boat lurched under the load and then slowly the mast came out of the water. It labored under the weight and I was sure it would break, but the boat shook and shuddered and tried to come back upright. The wind kept slamming us over. The water around me was tinged red, and I looked down at the bare bone of my thumb. The knife had cleanly slit it open and the skin peeled back neatly to reveal a stark white bone.

"Oh damn, now what?" I muttered to myself. I could not feel any pain, but the blood was making a mess of an already messy situation. I was skidding around trying to brace myself so that I could finish removing the cover. Inch by inch I cut away at it, and then finally it was gone. I wrapped the runner tail around a winch and opened the jammer. The load was huge with the full weight of the mainsail, battens* and boom pinned

---

* **Battens** - carbon-fiber rods that are used to support the mainsail and make it easier to handle. There are six on *Great Circle*.

up against it, but I felt the pressure ease as I paid the runner out. The boat slowly came upright and the sail started to flap and flog in the wind. I saw that the battens were broken and protruding from the sail like the exposed bone on a broken leg. The mainsail flogging was tearing at the mast; the whole rig was shaking and shuddering. My nerves were screaming for relief.

"Okay, okay Hancock," I said to myself. "Just get the damn thing down and you can get out of here." I climbed below again and checked the chart. We were still drifting towards land now less than four miles away. "Jesus," I muttered. "A stinking lee shore in a full gale without an engine." Blood was still pouring from my hand and I felt light-headed. I wondered if it was from lack of blood, or from a massive rush of adrenaline? Either way I felt like puking. The bulkhead was smeared with blood and my foul weather gear was crimson.

"Come on Eric, help me," I yelled. The sky above was black and angry. I wrapped the main halyard around a winch, carefully flaking it so that it would run out, and then eased the jammer. I felt the boat shake as the load came off the sail, and saw the luff drop a few feet. I eased the halyard again and the sail dropped another foot. I took a wrap off the winch and paid out some more, but this time the sail did not come down. I was worried that the pressure of the wind would cause too much friction on the slides*, and I would have to grind the sail down. The boat was still rounded up into the wind and I didn't dare run off because doing so would have spun us in the direction of land. Instead I bore off a few degrees, gained some speed, and then shot back into the wind. The bow slammed into the waves

* **Slides** - nyletron (a kind of plastic) fittings that attach the sail to the mast. There are also ball-bearing cars used at the batten ends.

and the main flogged violently, but the sail would not come down.

"Now what," I yelled. The halyard was free and the top ten feet of the sail was loose, but it appeared as if a slide was hanging up. It was hard to see in the fading light, but I was sure that the slide had caught on something. I wrapped the halyard around the winch again and ground it back up a few feet. The rain and gray skies had visibility reduced to zero. From the base of the mast I looked aloft and could just make out the head of a bolt sticking out from the mainsail track.

"Oh man, now I am stuffed." It was definitely the head of a bolt. The track is bolted to the mast and with all the shaking and banging one of them had backed out. The head was projecting less than a quarter of an inch, but that was enough to stop any of the slides from passing by.

For a moment my legs felt weak, and I slumped onto the cockpit sole to think. The rig was still in the boat, the boat was upright and despite the flogging and broken battens, the mainsail seemed to be in reasonable condition. The only major problem was land to leeward. Hard rocky land. I have never had to call for help and I was reluctant to do so now, but what were my options? If I could not drop the sail I would drift slowly towards the beach and sometime in the middle of the night we would be wrecked. I had no energy to do anything. Blood was still pooling on the deck mixing with the salt water, tingeing everything a soft crimson.

My other worry was insurance, or more to the point, lack of insurance. Every single asset I had accumulated in forty years of living was right under me, bobbing towards the beach, and I was not sure how to save it, or for that matter, how to save myself. The insurance was paid in full for coastal sailing, and sailing with crew, but sailing alone is viewed as dangerous and abnormal behavior and the insurance quote was equal to the price of a small car. I had no safety net. I was going to have to save my own skin.

Spurred by the thought of hitting the bricks, I went back to the mast and looked up. It was almost dark and the rain and wind whipped through the rigging. I could no longer see the bolt protruding from the track, but the sail was still hung up and I knew that one way or another, I was going to have to get it down. Back in the cockpit I loaded the inboard reef line onto a winch and started to grind. The wraps snugged up on the drum and the reef line went taut. I kept on winding, feeling the load, desperate to get the sail down. There was loud bang and bits of black nyletron from the disintegrating slide fell to the deck. The downhaul went slack and the sail dropped a few feet, and then stopped. Next slide. This one was a ball-bearing car at the inboard end of a batten. I felt the slide snug up against the bolt, and wound the downhaul. The reef line was strung taut like a guitar string, but still the slide held fast. I jammed the downhaul and re-led the line to my biggest winch. With all my weight powered by a spurt of adrenaline, I cranked the handle. There was a ripping sound, then a shower of ball-bearings. The downhaul went slack and the sail dropped another few feet. Another slide, another grind and another shower of nyletron.

It was completely dark by the time the sail was down. I took a knife to the last few feet and lashed it to the boom. It was strangely quiet without the flogging. Below deck the lights cast a pale glow on the interior. It was a mess. Spaghetti sauce everywhere, and the bulkhead smeared with dried blood. I looked at my thumb and the wound lay open. It had stopped bleeding, and for the first time since cutting myself I started to feel the pain. A slow throb at first, followed by a searing hot pain running up my arm. First the storm jib, and then I would bandage the cut. We were still drifting towards Faial, now just over two miles to leeward. Its mass was clearly visible on the radar screen. From the deck I could make out an occasional light on land.

The storm jib was buried under a pile of sails and spilt vegetables. I found the bag and threw it out the fore-hatch, followed by stronger jib sheets. Hanking the sail on was easy compared to the rest of the afternoon's activities, and in a few minutes I had it up and drawing. I eased the helm down, bore away and gybed. For the first time in four hours my course was leading me away from land. Away from danger, but also away from rest. With the autopilot broken, I would have to remain at the wheel until we docked in Horta. I lashed the helm riding a course away from the island, and went below to tend to my hand and make some soup. My body was shivering violently despite the warm night air. I wished there was rum on board, but we had finished the last bottle in Flores. Maybe it was better that way, I thought. The rum would only dull my senses at a time when I needed them most. I found a bandage, wrapped it around my thumb and tried to decide what to do next.

My first impulse was to sail as far away from land as possible, away from danger, but getting back without a mainsail would be a problem. *Great Circle* does not sail very well to windward with only a jib. My batteries were low and I was tired. The autopilot was broken, and my sails were ripped. Making Horta would be a relief, but sailing along the coast at night with only a storm jib was not smart. I was close enough to see lights clearly despite the rain. I gybed again and aimed at the narrow strip of water between Faial and the big island of Pico. My course was a close reach to the channel, and then a short beat up to the harbor of Horta. "Go for it," I thought. If something goes wrong I could always bail out and get clear of land. I was too tired to make rational choices, and the pull of land was strong.

Fortified by the soup and the thought of a safe harbor, I

I circled once to be sure it was a good place to dock, and then tied up alongside.

sheeted the storm jib and braced myself at the wheel. My speed over the ground was less than two knots and would diminish as we encountered an adverse current between the islands. *Great Circle* hobbled along the coast, short tacking through a wide arc and making slow progress towards Horta. Sometime after midnight we were abeam of the small airport half way up the coast, and by four in the morning I could make out the loom of the city and see the lighthouse on the breakwall. My hand throbbed and my body felt like a soft sponge that had been beaten with a stick.

It was only the persistent rain that kept me awake.

The faint fingers of a gray dawn were filtering through rain squalls as I made the breakwater. My outboard started on the first pull and I dropped the storm jib on deck. It had been nineteen years since I was last in the Azores and my memory of the place had faded. I recalled that there was a big seawall where all transient sailors painted their names, and the name of their yachts. I had left my mark there aboard *Battlecry* on a passage between the West Indies and England, and wondered if it would still be there. A new inner harbor had been built and I could see a forest of masts. The old seawall was empty except for some of the local fishing fleet and a small freighter unloading cargo. I circled once to be sure it was a good place to dock, and then tied up alongside. I felt exhaustion seep through my body. A lee shore in a full gale with no insurance was every sailor's nightmare.

And then I remembered Tabarly.

The squalls were moving out to sea and the sky to the west was clearing. The sun hit the whitewashed buildings, and the green fields above Horta were rinsed clean, their fragrance carrying over to me. I could smell cows and hear sheep bleating in the distance. Down below was a mess, but I crawled into my bunk, flicked the light off and dreamed of a cabin in the mountains, far, far away from the ocean.

A DATE WITH A WALL

"Enjoy yourself to the brim and remember that whenever you get into trouble, trouble is only temporary."

Shelly Barron

I watched the tall peak of Pico slip silently into the haze behind me. For a while I could make out a dim outline, and then it was gone. I was alone again. Bermuda was two thousand miles away and would be my next landfall, but between my bow and that distant island lay open ocean. I said a silent prayer for a safe passage, and then set about organizing my stores and plotting a course. The past few days had been busy and I was glad to be off. I listened for the slap, slap of the waves on the hull and quickly settled into the rhythm of life on board.

My stay in Horta had been hectic. There was much to do to repair the damage done by the knockdown, and a team of volunteers lent a helping hand. The mainsail was my biggest concern. It had taken a pounding and needed a lot of attention. It was now patched and I hoped it would hold together until Bermuda. The autopilot head had been replaced and was

Other than the heat, my trip to Bermuda was a perfect ten.

working fine. I could hear its moan as the hydraulic arm tugged at the steering quadrant, and I felt the boat respond to the course dialed into the computer. *Great Circle* heeled to the breeze and sliced through the clear water.

It felt good to be back at sea again.

As the day faded into night I sat alone on the deck and watched the water trickle in my wake. I had set a course that would take us south for three hundred miles, before slowly turning to the west. I needed to skirt the Azores High, and a direct course for Bermuda would have taken me straight through the windless zone. So far my tactics seemed to be working. The wind was on my beam and blew warm and soft. It was the same temperature as my skin and I could not feel where my skin ended and the air began. In the east a full moon rose slowly from the ocean. It shed its silver light onto the dancing water and sparkled in my wake.

It was a night of pure magic.

And then suddenly I remembered. I had forgotten to paint my name on the wall. It has become tradition for all sailors who pass through Horta to paint the wall with elaborate graphics. The old seawall, the new seawall, the docks and the jetty are all covered with paintings, and it is becoming difficult to find a place to leave your mark. The day after I arrived in Horta I searched for the painting I had left behind in 1979. It took hours to find and was faded and chipped, but the red spinnaker with a white cross was still visible, as was the name *Battlecry*, painted below. I had wanted to leave *Great Circle*'s logo for all to see, but in my haste to leave, I had forgotten. While I am not superstitious, legend has it that it is bad luck to leave Horta without leaving your mark. I gazed at the water rushing by and hoped that my luck would hold, at least until I made it safely across the Atlantic.

I remained awake all night keeping a lookout for ships, and as dawn lit the sky I crept into my bunk and slipped into a deep sleep. My radar was alarmed to warn me if anything came within range, and I had set my alarm clock to roust me in an hour. I hoped shipping was keeping a good lookout.

The sun rose early. It pounded down on the boat and I woke in a pool of sweat. I lay in my bunk thinking about the day. It was going to be hot. It was already too hot to be below so I climbed onto my nav seat and dug through the nav station looking for sunscreen, but could not find any. I searched my stores. No luck. I searched everywhere and eventually gave up. I knew that I had brought some with me, but could not find it. Two weeks in the tropics and no sun protection – I hoped I could find the toilet paper.

The wind had swung aft with the new day, and I dropped the big genoa and set a spinnaker. We were already two hundred miles from Horta and I was enjoying the passage.

***

A few days after my arrival in the Azores I returned to the US. The knockdown had banged an important lesson into my brain – you can never let your guard down when you are alone. The ocean is forgiving to a point, but push your luck and you pay dire consequences. I had been lucky, but then I was born under a lucky star. I left the boat on a mooring in Horta harbor and flew home to get married; and in doing so embarked upon another voyage as tumultuous and unpredictable as any transatlantic crossing. My honeymoon would be a solo passage back across the Atlantic.

A week after returning to the States I woke in the middle of the night in a cold sweat. I had been dreaming that the dockmaster in Horta was calling me and telling me to come and get my boat. I lay quietly with my heart pounding trying to recall my dream. The dockmaster had been yelling into the phone that there was danger coming. I had to move my boat, he said. I glanced at the clock next to my bed and saw that it was two in the morning. After a long while I went back to sleep. It was not restful and when the sun came up I opened the curtains and flicked the television on. The lead story was about an earthquake that had hit the Azores. The town of Horta had been woken just after dawn to a tremor that destroyed much of the island. The quake had happened at two in the morning, US east coast time.

When I returned to the Azores I was amazed to see the damage. One third of the rural buildings were demolished. Roads were damaged and bridges had collapsed. There was not much damage in Horta, and people in the marina had only felt their rigging shake. They said it sounded like a low-flying jet passing overhead. *Great Circle* was on a mooring; and safe. I felt that luck was running my way and dismissed the sunscreen as a simple oversight – until I tried to start the generator. It was dead. I bled the system and cranked the engine; still nothing. I changed the fuel pump and tried one more time; still dead. It would take me over a week to figure out the problem until which time I ran on no power. I shut off all lights, alarms, instruments and stereo, and used the solar panels to power the autopilot.

I started to wish I had painted my name on the wall.

The days blended into nights and back into days as we slowly inched our way across the chart. The wind remained aft of the beam, and other than the intense heat and lack of sun protection, it was turning out to be a fine passage. I caught a few fish and made sushi. The Portuguese cheese and olives were good, and a healthy dose of fresh air made a world of difference to my temperament. The knockdown off Horta was fading into a distant memory. It took me a week to get the generator working again, and once I had the batteries charged I fired up the stereo and plugged in a disk, but soon shut it off. The silence had become addictive and the commercial sounds of my domestic life seemed intrusive. I preferred the

noise of water rushing by the hull, and wind whistling in the rigging. All was right with the world except that there was no sea life. Nothing. No birds, no dolphins, no whales, no flying fish – nothing. Every so often I would see some piece of garbage float past, but that was all. I was disgusted by the pollution and saddened by the lack of living things.

After two weeks we approached land.

My chart showed a thin line snaking towards a small island far out in the Atlantic. Bermuda was drawing closer, and suddenly we were there. I made landfall on St. George's light and sailed right into the harbor. There were a few boats at anchor, but most had headed for safer waters with hurricane season fast approaching. I tied up alongside the customs building and felt a deep sense of accomplishment. I had been lucky with the weather and guessed that other than misplacing the sunscreen and generator problems, my luck was holding. I still wished that I had painted the wall. My luck was holding, but it was soon to run out.

***

I left Bermuda a month later – coincidentally at the same time as hurricane Bonnie was flirting with the Gulf Stream east of Miami. Forecasters were predicting that the storm would either track across Florida, or turn north and pass between the US mainland and Bermuda. It was a few days from my location and I reasoned that I would have plenty of time to turn around and go back if the storm came my way. I was anxious to get back to the United States, and cast my lines ashore.

I sailed out of St. George's harbor just before midnight, and by the following morning was one hundred miles closer to Marblehead. Bonnie had stalled and was sucking up the warm waters of the Gulf Stream, and intensifying. I was getting nervous. It had become a Category Five hurricane — the deadliest kind. The storm was packing one hundred mile-an-hour winds and there was a distinct possibility it might head north. There was no use pressing my luck. I had pushed it far enough just by leaving so I turned *Great Circle* around and headed back to Bermuda. The afternoon air was still and moist, and the wind died away to a passing zephyr. I motored, and as the sun dropped into the ocean behind I grilled a steak on the barbecue. It was a spectacular sunset – I hoped to be back on land before midnight.

Approaching Bermuda from the west is much trickier than it is coming in from the east. The northwest side of the island is bordered by a huge reef, but it is well lit. I called Bermuda Harbor Radio and asked for the coordinates of the lights on the reef and noted them in my log. My chart was for the entire Atlantic Ocean and did not show details of the approach to Bermuda. I programmed my GPS and made landfall on the first light at two minutes past ten. The lights of Bermuda twinkled in the distance and St. George's lighthouse cast its beam far out to sea. I sailed right up to the light so that I could check my GPS, and then altered course for the second light.

It was less than a mile away.

A dim red glow from the night-lights illuminated the interior and the green radar screen showed a mass of land off to starboard. I could see the light we were heading towards on my screen, directly ahead. The waves slapped their familiar rhythm on the hull as we heeled gently in the dying breeze.

And then we hit.

The noise of carbon being wrenched from the hull by the sharp edges of coral still ring in my head. The tracking fin hit first and ripped out of the boat leaving a gaping hole. I felt the keel hit as I ran up the companionway and grabbed the wheel. I swung the boat away from land and we hit again.

This time I could feel the rudder jerk in my hands. It was a dark night and the water was still. We had been sailing at four knots on a calm sea. There was nothing around us. Ahead I could see the light we had been sailing towards, and behind the dim glow of the one we had just come from. Frantic, I gybed the boat in towards the reef and we spun in a tight circle. The boom came across and miraculously we found open water and I sailed away from danger. Below, the water was flowing swiftly. It was already up to just below my knees and filling fast. I slammed the watertight bulkhead closed to contain the leak in the forepeak, and started the bilge pump. *Great Circle* wallowed wounded and I wept for my stupidity. Later when I looked at a chart I saw that the reef sticks out beyond the straight line. You cannot sail directly between the lights.

It was nearly a fatal error.

With the watertight bulkhead containing the leak, the forepeak filled until the level inside matched the sea outside, and then it stopped. I dropped my sails and started my outboard, and once I was certain that we were not going to sink, I motored slowly, and painfully, back to Bermuda. My dream of sailing around the world alone had come to an end, and my respect for superstition and legend had grown.

I vowed to return to the Azores to paint my name on the wall.

***

The following winter was one of the worst of my life. I floundered without direction struggling to shake a mood of melancholy that had settled on me. I repaired *Great Circle*, but it was too late to make the start of the Around Alone, and I watched with dismay and a slight depression as the fleet sailed from Charleston towards South Africa. I had worked for two years to be a part of that great adventure, but it was not to be. I repaired the boat and it was soon back sailing – I remained afloat in a sea of self-doubt. My pride was wounded and my psyche damaged. I kept wondering if the wall had something to do with it. It was only a painting that I had forgotten. How could that simple oversight cause so much trouble in my otherwise charmed life?

By spring the Around Alone fleet was already on its last leg back to the US. Cape Horn and the Southern Ocean was behind them and they were headed for the finish and fame. I was still floundering, but had decided that it was time to return to the Azores to splash some paint on concrete. It had to be done. I heard of a boat that needed a hand for a trip from Newport to Horta, and signed on for the passage. There would be two of us on board, and we would sail a well-appointed forty-six foot yacht back across the Atlantic. It was a nice boat but it too resembled floating furniture, as I like to call yachts that look like a bungalow in House Beautiful. I barely gave a moment's thought to the trip other than to pack some paints, paintbrush and a toothbrush, and think about what kind of design I would do when we got there.

We left Newport on a starlit night in early June.

By the second day we were pounding into steep seas and strong headwinds. I looked at the pilot chart for the North Atlantic and it showed a two percent chance of easterly winds for the month; none of them should blow more than 12 knots. The strange strong wind continued for another day and then died. I was glad to get my two percent out of the way early in the passage.

And then the trouble started.

My shipmate woke me in the night to announce that he had dropped the fuel separator for the main engine overboard, and did not have a spare. It was potentially a serious problem, but we found a plastic glass in the galley that could be modified to fit the fuel filter. We epoxied the two and set the jury-rig aside until morning. I wanted to be sure that the epoxy set properly before we ran diesel through it. We would run on the power we had and use the generator to charge batteries in the morning, but something was draining them, and before dawn they were flat and useless. We cranked up the generator, but it would not charge. My bad luck with engines was returning and by noon that day we did not have enough power to light up the compass light – and that was the least of our problems.

Modern sailboats are designed to consume electricity and provide all the conveniences of home. The winches turn effortlessly at the push of a button. The autopilot keeps the boat on a course straight and true. The drinks are cooled with crisp chunks of ice, and the frozen steaks are thawed before grilling. Lose power and you lose it all. I was most concerned about having to hand steer the rest of the way. We took up a watch system of helming two hours on, and two off, twenty-four hours a day, and spent the off-time trying to bypass the systems on the boat that needed power to operate them. The gas stove has sensors to make them safe, but no power means no sensor, and no sensor means no gas. There was no running water, no lights and worst of all, the food went bad. And then the wind swung back into the east and a gale started blowing. I thought long and hard about superstition and legend and we talked about turning back, but I was adamant about pressing on – I had a date with a wall.

The wind gusted over forty knots and the seas on the edge of the Gulf Stream were big and confused. We were able to rig the helm so that the boat sailed itself while we sat below and watched as the waves crashed over the deck. We could feel an occasional weightlessness as the boat plunged from wave crest to bottomless trough, but the boat soldiered on and we were safe and dry.

Until the boom separated from the mast.

The fitting that attaches the boom to the mast had ripped in two and the reefed mainsail was flogging frantically while the aluminum boom thrashed about trying to wrench the rig to pieces. We managed to lash it and effect a temporary repair. The boat wallowed and lurched from gust to gust as sheets of rain blew horizontal with spindrift lacing the gray water. There was no sign of an end to the storm, and as I made my way below I felt my back let go. There was a sharp pain in my lower back and then a searing hot pain throughout my body.

My battle with back pain was back.

I made it to a quarterberth and wedged myself in with cushions, trying vainly to lie still. Any movement of the boat was painful, and in an easterly gale there was a lot of movement. I gazed blankly at the overhead and wondered what else we were in for. Two days passed before the wind subsided, along with the pain. I crept tenderly behind the wheel to steer while my shipmate gratefully found his bunk – he had been at it for forty-eight hours without a break.

After a week at sea the wind slowly came aft and the sun came out. All we needed was power, and life would be back to normal. Well, some food would have been nice. We were reduced to eating pasta and cans of tuna, but at least we were not starving. If we lost the wind completely we would be in trouble. Life settled into a rhythm. Families back home became increasingly worried. There was nothing we could do other than plod on towards Horta.

And then the wind died.

We were twelve days into the trip and the ocean was flat and featureless. It did not resemble the sea we had been looking at for so long. The steep gray waves were now gently rolling turquoise. The sails slatted incessantly in the dying wind and filled with the occasional zephyr, but mostly we drifted in circles while our ETA on the hand-held GPS showed us arriving in Horta in mid September. I wondered how much tuna we had on board and guessed it would not be enough.

As the sun set on our thirteenth day, dolphins joined us to watch the spectacle. The air was heavy with squalls reflecting the evening light. I watched a small rain cloud on our beam and thought I noticed a hard edge to it. The edge grew more distinct until the bridge of a ship appeared. A ship with food and perhaps an emergency battery charger. We had saved the hand-held VHF radio for just such an emergency, and a scratchy call had us in touch with the *Crown Topaz*, a Chinese freighter bound for Miami. The captain did not speak much English, but for some reason he did understand the words "jump-start." Perhaps he drove an old car. Whatever the case I saw them alter course and head in our direction.

We inflated the dinghy and got ready to launch it. I called the captain again and explained that I would be in the dinghy and they should drop a battery down to me. I did not think it was a good idea to have the ship come alongside, and the captain agreed, or at least he seemed to. As the light faded I set sail and rowed in the direction of the ship. Our boat bobbed in the slop and looked small and vulnerable. The ship was drifting slowly and I intercepted it at the bow. From my vantage it looked massive. The sides were sheer and steep steel. I scraped down the hull fending off with an oar, and halfway down looked up at the smiling faces of the Chinese crew. They were dangling a battery at the end of a line attempting to drop it into my boat. I thought about what might happen if fifty pounds of battery landed in my rubber raft and hoped they were kidding. They weren't, but fortunately the ship was moving too fast. I skidded past the drop-site ducking to miss the battery, and drifted aft towards the propeller. I was desperately attempting to row away from the ship, but the wash sucked me back in and I was headed right for the prop. By luck, or by chance, the captain shut the engine down, and as I was spat out the back I passed the huge propeller missing it by two feet. The blades were motionless — they were knife sharp and reflecting dully in the fast fading light. The ship kept on drifting. I was left bobbing in its wake. The boat was lost in the dusk and squalls.

I spun the dinghy around and started to row.

In the distance I could just make out the ship maneuvering alongside the boat, and my shipmate vainly attempting to hang fenders over the side. They were a long way off and our yacht looked pathetic and small alongside the freighter. Even the freighter was dwarfed by the vast ocean. I pulled on the oars and slowly gained on them until I could see the battery on deck and a small Chinese man scrambling down a rope ladder. The crew on the freighter rigged spotlights to help out and the boat rocked gently in the lee of the ship. My shipmate and the engineer were below fiddling with jumper cables and the battery. I scrambled on board in time to see a garbage bag of food land on the foredeck, followed by another filled with drinks. I gave a thumbs up as the engine roared to life, and watched as the small engineer climbed back on board his ship. The *Crown Topaz* pulled away and sailed into the dark night. Her stern light faded with the arrival of a slim moon. The spirit of seamen was alive and well and I thanked the captain and thanked the gods. We would make Horta before September.

Then the wind picked up and swung back into the east.

We beat through the night eating cheese sandwiches, and listened to the sound of the autopilot groaning back aft. The next morning the wind had picked up to gale force and we slammed and slogged towards Horta. I thought of my trip across a year before – it might as well have been a different ocean. The dolphins had given us luck, but it was paint on the wall that would change it for good.

Horta was still two days hard sailing away.

Dawn on day sixteen broke wet and windy. The green hills of Faial were laced with lashing rain and the tall peak of Pico was shrouded in a low fog. The wind funneled between the two islands as we plowed our way up the coast. The weather was not going to give us a break. I remembered creeping towards Horta a year before with ripped sails and a heavy heart, and thought about all that had happened since. It had been an eventful year, not all good, most of it interesting, some of it challenging. It was time to paint the wall and head home. My plane was scheduled to leave for Lisbon right after lunch. We tied up alongside and I felt a weight lift from my shoulders. There was just enough time to clean the boat, lunch at Café Sport, and a bit of painting.

The wind died and the sun came out drying the concrete pavement. I found a small spot without a boat name, and left my mark. It said *Great Circle*, 1998, and in the top right hand corner, I wrote my name.

UNDER AFRICAN SKIES

*"I learned what every dreaming child needs
to know – that no horizon is so far you cannot get above it or beyond it."*

Beryl Markham

I grew up at the bottom of Africa in a small town with a big name. A juxtaposition of the last names of the two men who first settled there. They named their settlement; Pietermaritzburg.

The town is nestled among the rolling hills of the Natal Province of South Africa, midway between the Drakensberg Mountains and the Indian Ocean. It used to be an orderly town with soft edges, but the soft edges are gone now, eroded by time and the turbulent changes that swept the country over the past two decades. Where the hillsides were once green and covered with wild flowers, they are now red mud, laid bare from neglect and overgrazing. Even the smell is different. The fragrance of jacaranda blossoms has given way to car exhaust fumes and cooking fires. The air of British colonialism that permeated all parts of life has gone, replaced by the rough and tumble of a struggling Third World city. The town has slowly grown to fit its name.

When I was small, the streets were safe and the town claimed the unofficial title of being the last outpost of the British Empire. At times it seemed more English than England, as expatriates clung to traditions long since given up back home, but treasured as fond memories in Africa. During the summer we dressed in white and played cricket, while our parents also dressed in white and went lawn bowling. In the winter it was rugby, whether we liked it or not; and I did not. I was too small to be of any use on the field, and not much of a team player. I was more interested in boats. I longed for the weekends when I could go sailing. I longed to leave the routine of school and work and slip away to the familiar surroundings of the waterfront. The smell and sound of the lake immediately erased all the muddle and mess and stress of growing up. The moment I set sail it was washed away and replaced by the slap slap of clear water on the hull. I knew from an early age that my life would be tied to the water, and early seeds of adventure were planted in my resolve.

My earliest memory is of a short flight. I was six. I remember the take off, and I remember the landing. The bit in-between has gone, but then it was a very short flight – less than a second, actually. I launched myself off the verandah wall with a vague belief that a positive attitude and a bit of luck would see me safely down. I had makeshift wings strapped to my arms. The second my feet left the wall I discovered the undeniable effect of the pull of gravity, and dropped like a stone. Young bones are supposed to bend and mine did at first, but everything has a breaking point as my left leg discovered a split second after making contact with the hard earth. I don't remember much beyond the landing. I recall the sterile smell of disinfectant burning my nostrils while I waited for the cast to be molded, and I remember being given the heel-guard to keep as a souvenir when the cast was cut off three weeks later. I had healed quickly and was sent back into the world to see what new damage I could inflict upon myself.

It didn't take long.

During the winter we would visit my grandmother for Sunday lunch. You could smell the roast beef from the parking lot. With ice cream and chocolate sauce for dessert, it was worth getting dressed in our best clothes and sitting through a few hours of adult conversation. One Sunday, not long after I had regained the full use of my broken leg, I ran headlong into the back of my father's car using the top of my head as a brake. I do not remember much beyond the initial impact. I do know that I ruined my best clothes and lunch at grandma's, and recall that now familiar smell of disinfectant.

Some years passed before I inflicted more damage. Like a lot of families, we did not own a swimming pool and instead would visit the public pool. I decided that diving was going to be my forte, and spent the morning perfecting my somersaults and belly-flops. By lunchtime I had them mastered, and decided to seek out new challenges. A simple back-flip. The dive would stun my group of admirers and I would be able to claim my station as the diving expert of the pool. I positioned myself on the edge facing away from the water, and shuffled backwards until my toes were clinging to the lip. My heels hung over the water and I balanced for a moment building up courage. I teetered for a second, and then with peer pressure coaxing me on, I leapt up into the air. The moment my feet left the ground I knew that it was going to be the dive to beat all dives. I sprung upwards, arched my back gracefully and crash landed right where my feet had just been. For a split second I teetered on the edge, my head flattened against the hard cement, and then in slow

Growing up in South Africa – at age five.

motion collapsed into the pool in a bloody heap. I remember looking back into the water and seeing a pool of crimson blood slowly mingling with the turquoise water. I remember the disinfectant in the emergency room, and the return trip to remove the stitches two weeks later.

I survived my early years albeit with a few scars, and when I turned seven I was scrubbed and polished and sent to school. I joined the ranks of other identical-looking scrubbed and polished children all sent to absorb the wit and wisdom of identical-looking scrubbed and polished teachers. The schools in South Africa in the 1960's were segregated, not only along racial lines, but along gender and language lines as well. I received a pasteurized education, while attending an English-speaking, whites-only, all-boys school. It seemed perfectly normal to me at the time. Children accept their parents' ways without questioning the logic of it all until later in life. It took me twenty years before I figured there might be a better way. Each day, as I rode my bicycle to school, I passed black children heading in the other direction to their schools in the country. They were all walking, most of them barefoot, and it never occurred to me that there was anything out of place. They had their schools, we had ours and that was that. My biggest concern was not puncturing my tire.

I would arrive at the school gate promptly at 8 a.m. and wheel my bicycle to park it neatly alongside the other bikes. Morning assembly began at 8:01. Attendance was compulsory. Most of us made it in the nick of time on the threat of detention after school, or a good caning if the problem persisted. The entire school would assemble for morning prayers recited in Latin, while our Latin teacher beamed down on the assembly. He was the only one that had any idea what was being said. His name was Joe Harding, but we referred to him simply as "Spud." Sitting alongside "Spud" in the teachers' gallery, was "Meatball," our aptly-named geography teacher who never anticipated that I would put my lessons to such good use. Neither did "Howie," our English teacher who glided from class to class like a sail-boat with a list.

"You must read more books," he constantly admonished. "Books are the source of all knowledge." I took his advice to heart and read them by the dozen, mostly about travel and adventure. They planted a restless wanderlust in my blood.

The vice-principal was simply known as "Max" – that was his real name, although it might well have been short for "Mad Max." It was rumored that he had a steel plate in his head; the result of an accident received during the second of the big wars. On hot days the plate would expand causing headaches that immediately translated into a bad temper. "Max" and the headmaster were allowed to cane students and we prayed for cool weather if "Max" was handing out the punishment. Most of us paid a visit to his office at least once.

Morning assembly was a boring affair with the headmaster interspersing his comments between the unintelligible prayer and hymns sung out-of-key by an overanxious choir. The only change in routine was on Monday mornings when the sports results were read – and the mood on campus was set for the week by how well the teams had performed. The headmaster's usual routine was to stride purposefully into the hall after everyone was seated, turn and glare down at us from behind his podium, glance towards the singing teacher who would strike up the choir, and then seat himself front and center to conduct the proceedings. On one particular morning sabotage had taken place. During the night someone had sawn most of the way through the back legs of his chair, and as he took his seat, they gave way. Those who laughed were later singled out for a good caning. The headmaster meanwhile found a new chair and continued without comment. The choir seemed more off-key than ever. An inquisition was held, but they never found the culprit or the tool used in the crime. We were all made to stand on the rugby field in the blazing sun until someone either owned up or forwarded incriminating evidence. I had no idea who did the sawing, but he remains a hero in my mind three decades later. It takes nerve, cunning and careful planning to pull off sabotage, and I envied his skills.

During high school, someone suggested a feat that, if performed, could earn you a "Victoria Cross." The highest award for bravery on the battlefield, was to be our reward if we accomplished the task. It wouldn't be easy. You had to run naked from our school, along a main road for over a mile, to the all-girls school at the other end. When you got there you had to climb to the top of the high-diving board, jump into the pool, swim twenty-five yards across, and then run back to our school. Get caught and there was no VC. It seemed like a perfect challenge and I was among the first group to go for it.

Five of us gathered behind the changing rooms. The evening air was tepid and damp; a perfect night for a run, but each of us was hoping that one of the others would call the whole thing off. It had seemed like a good idea during the light of day, but darkness exposed hidden doubts. There had to be some reason for bailing out, but everyone was silent. No one was going to be the first to chicken out and peer pressure has forced worse judgment calls. Two students had already been dispatched. They would be waiting for us at the girls' school to ensure that we carried out that end of the deal. Another two acted as witnesses for the start, and hopefully, the finish. It was a little after 9 p.m. I would have preferred to run at midnight when there would be less traffic on the road, but there was no way I would be able to be out after midnight on a school night. As it was, I had to make up some story about doing homework over at a friend's house. The changing rooms were locked, but it didn't matter. We did not need the privacy they offered when our task was to spend the next hour running naked in public.

We slipped our clothes off, feeling vulnerable in the moonlight and awkward in our puberty. The first obstacle would be the night watchman. We had to pass his post to get through the school gates, and he would surely blow the whistle if he saw us. We crept from bush to bush hiding from the shadows that instantly took on the form of a teacher. In a few minutes we made it to the gate and peered out into the street. No sign of the watchman. There was a single car coming towards us; after that it was all clear. We waited for the car to pass before making our move. We would run the main road for over a mile, and there were only a few bushes along the way where we could hide. The trick was going to be finding a bush when we needed one, especially one that could shelter five naked boys.

The car passed with the driver oblivious to the moving shadows. My heart was pounding. Getting caught would mean a severe caning by the headmaster, and perhaps even expulsion from school. Still, what was a VC worth if there was no risk involved? We were about to start running when another car appeared.

"Wait," I said to the others. "Just let this one go by and then we'll make a run for it." The car slowed as it approached the gates, and then suddenly it turned onto the school grounds. We shrank back into the shadows keeping very still. The car stopped and I was certain that we had been seen.

"No one move," someone whispered. I felt the urge to run for it. They would never know it was me. I could get back to the changing rooms, grab my clothes and be out of there in a moment, but peer pressure kept me frozen to the spot. I heard the car's engine rev, and it pulled forward slowly. As it slid by our hiding place we could make out the familiar outline of one of our teachers. He was not looking our way.

"OK guys, let's make a run for it." The pent-up adrenaline squirted directly into my bloodstream, and I sprinted as fast as I could out into the street and up the road. My legs were a blur as I tried to make as much distance as I could before the next car appeared, but the others were much faster, and I was quickly left behind. That was the last thing that I needed. Running in a pack was one thing. You had the security of others to hide behind. Running all alone was a frightening experience. I felt very exposed and very naked. The driver didn't see us and we chased the flickering taillights up the street.

Less than a quarter of the way along another car appeared. We scrambled for a bush and waited until it had passed. The road was dark again except for the dim glow of the street lights. Another couple of cars appeared and we crouched lower, hoping to become as inconspicuous as possible. The street seemed busier than it had all day. We hid in silence hoping that the collective thumping of our hearts would not give us away. The cars passed, the drivers oblivious to the huge feat that was taking place just feet from them. Once again the street was dark and empty, and we ran on.

The worst part was running beneath the street lights where we were instantly bathed in light. What if we were caught? How would I explain this to my parents? Too late now. We found the fence surrounding the girls' school, scrambled through a hole and melted into the shadows. Laughter spilled out nervously.

"We made it this far. There's no going back now," I blurted, quickly regretting that I had spoken. It came out sounding as if turning back was something I had been contemplating, and even though it was, there was no way I was going to admit it.

"The school is all dark," someone noted. Through the trees we could make out the buildings and they all appeared to be unlit. The swimming pool was on the near-side of the school, but it was close to one of the girls' dormitories, and we would still have to make our way past a few other buildings before getting there.

"Shit, what's that over there?" Someone pointed in the direction of the nearest building. "It looks like a person." We all squinted, but could see nothing.

"Don't be a fool. There's no one there." I was certain that he was seeing things, but kept on staring into the darkness. All the trees started to look like teachers and night watchmen swaying their way towards us. My heart lurched as one of the shadows moved.

"Christ, guys, we are going to get caught," I said.

"Shhh, keep still," someone whispered, and then, "oh fuck, there *is* someone there." The shadow moved again, and this time we could all see a person clearly outlined against the building. Another shadow appeared alongside the first.

"They're coming this way." My breath caught and my throat felt dry and ached. Perspiration squirted onto my palms, and for a second time that night I fought the urge to run for it. The shadows moved slowly in our direction. We shrunk back hoping to become instantly inconspicuous.

"Hey you pussies, what are you doing just standing there? You are supposed to be swimming." It was the two witnesses that had been dispatched earlier to make sure that we did the jump and the swim. I took a breath and relaxed a little.

"Christ you bastards scared us," I said.

"It's OK. There is no one about. We've checked out everything for you. The night watchman is on the other side of the school probably locking up the classes and turning off lights." They looked awkward standing fully clothed alongside us. There is nothing like safety in numbers, even when you are naked.

"What time are the girls' supposed to have the lights out?"

"Nine-thirty and it's past that now. Anyway the place has been dark since we got here." We made our way slowly past the outer buildings. The girls' dormitory was dark and silent. It was on the far side of the swimming pool. I wondered if they would hear the splashing as we jumped into the water. By the time they heard the noise we would be across the pool, out the other end and well on our way. We crept forward. In a moment we would be completely exposed standing on the diving board, but for now we were still in the shadows. I looked up hoping for a cloud to pass in front of the moon, but it was a clear African night. We would just have to go for it.

"OK, guys this is it," I whispered. We climbed onto the ladder and quickly made it to the top, pausing to catch our breath before jumping. All of a sudden the lights went on and we were bathed in floodlight. From the girls' dorm shrieks of laughter and cheering erupted. We could not see beyond the glare of the floodlight and stood frozen like rabbits caught in a car's head-light.

"Let's see what you've got," a voice called out from behind the glare. "Show us your dicks." We were busted and naked in front of two hundred laughing girls. I jumped, and so did the others. We hit the water with a collective splash and struck out for the shallow end. I hoped like hell that I would not be recognized. My dick had shrunk into itself from the cold and nerves, and there was no way I wanted anyone to see that there was so little there. There wouldn't be any chance to explain that it was much bigger on a warm day. By the time I made it across the pool, the others, all being faster swimmers, were out and had run for the shadows. A huge cheer went up as I scrambled out of the water and then I heard a voice call out at the top of her lungs.

"It's Brian Hancock. Hey everybody, it's Hancock." I fled for the safety of the shadows with my ears burning, my face purple, and laughter ringing in my head. I found the others huddled by the hole in the fence. The witnesses were with them, laughing harder than the rest.

"You Bastards," I said, "you tipped them off, didn't you?" We were the only ones that didn't know it, but the school had been tipped off earlier in the afternoon, and by the time we were on the diving board, the dorm was practically bursting with anticipation.

We scrambled through the fence and set off down the main road. This time we were much more cavalier. We jogged together and didn't attempt to hide when the first car drove by. The driver honked his horn and waved, and we felt like kings who'd conquered the enemy.

Our clothes were where we had left them, and we dressed breathlessly. We had made it. No one mentioned that we could still get caught if word of our antics spread from the girls' school, to ours. If questioned, we would deny everything and the rest of our class would stand with us, just as we all had done for the saboteur who'd cut the chair legs.

That night I had my first wet dream.

I gained some stature among my classmates after running for the VC, and there was no fallout from the girls' school over my size, or lack thereof. As far as I knew no one ever found out about the run and we were spared a caning and the possibility of expulsion. The following summer I managed to push the boundaries once again, only this time I was to feel not only the full impact of being on the wrong side of "the establishment," but the bite of the rod as well.

***

We held a sit-in over school regulations that restricted the length of our hair. The protest was led by a fellow VC holder, and irrespective of how I felt about hair length, I was in with him on the protest. On the morning in question a few of us assembled on the rugby field and spread the word that we would not be going to morning assembly until our complaints were heard by the headmaster. Word soon spread and our ranks swelled, but the protest barely caused a ripple and the headmaster made no mention of it in his morning remarks. I had no firm views on hair length, and when the sit-in fizzled we capitulated thinking that was the end of it. The headmaster had his own ideas.

His sources exposed the ringleader and inner-circle, and we received a summons. Being called to the headmaster's office was not good news. He was the head of "the establishment," and knew how to use his power most effectively. His office was at the end of a very long corridor. To get there we had to pass the administrative offices, the secretaries' desks, the teachers' lunch room, his personal secretary and finally to his door. Portraits of past headmasters stared sternly at us from gilded frames. Four of us had been singled out as troublemakers and we were under no illusion that we had been summonsed to discuss our demands for longer hair.

We knocked, but there was no answer. From within we could hear muffled voices. We knocked again, but still no acknowledgment. We waited. Keeping us waiting was part of the strategy. My palms were dripping sweat. Suddenly the door flew open and a red-faced headmaster filled the space. He looked much bigger up close than he did from a distance.

"Ah," he said, "please come in." His manner seemed friendly, and for a fleeting moment I thought that we had been wrong to jump to conclusions. Maybe this was about easing hair restrictions. His office smelled of leather and perspiration, and was dominated by a huge walnut desk situated in the far corner. We stepped into the room and closed the door behind us.

"Please take a seat," he said waving us to a bench against the wall. We shuffled towards it and sat down. Out of the corner of my eye I could see the rack of bamboo canes mounted on the wall. I tried not to look at them, but they held my gaze.

"Hancock, I see that you are admiring my display case," the headmaster said, piercing me with his eyes.

"Oh no, sir."

"Don't worry," he said, "you will soon have the opportunity for a closer look." My heart skipped. I knew for sure that we were not there to discuss new hair regulations.

"Do any of you have anything to say?" he demanded from behind his desk where once again he looked his normal size.

"No sir," we replied in unison. He made some notes on a pad and then picked up the phone to make a call. We sat in silence while he discussed lunch plans with his wife. The bench was hard and I realized that the smell of perspiration was coming from me, and not from the room.

"Well then," he said, having allowed us to sweat for another few minutes, "I sentence you each to four of the best. Except you, Bassett. You will get six for being the ringleader." It was not an appropriate time to say anything. We just hoped that we could get it over with as quickly as possible. Now that we knew what we were in for, waiting for the caning would be unbearable.

"You will all report to my office tomorrow morning at 8:30." With that he dismissed us and with a wave of his hand sent us scurrying for the door. The hallway seemed to laugh at us as we ran past the secretaries and out into the fresh air.

"Fuck," said Bassett, "I don't want to wait until tomorrow. It's the waiting that's the worst." We all agreed, but we also knew that it was part of our punishment. The system had the power and the headmaster knew how to use it.

That night I kept waking up and looking at the clock, but time dragged. My mind scrambled for a way to cushion the caning. I'd heard of a student who had sewed soft leather onto the inside of the back of his shorts to add an extra layer of protection, but I did not have any leather. I thought perhaps I could stick something down my pants, but was worried about it being spotted and getting me into more trouble. By the time the soft light of dawn cast shadows on my bedroom wall I had given up on the idea and decided to face the caning like a man. That was how I felt at dawn.

It would likely not be the same at 8 a.m.

It was a crisp, beautiful morning, but I felt like a prisoner about to face the gallows. Getting four of the best was a lot. Two was considered reasonable punishment. Six was the most you could get. I wondered how Bassett was feeling.

I was early for assembly for the first time in my school life. Everyone crowded around anxious to hear what we were going to do, but there was nothing we could do. Just sit through the assembly and hope for the best. At 8:29 a.m. we made the long walk back down the hallway to the office at the end. We were let in right away. The headmaster seemed to be enjoying our discomfort and asked us if we had anything to say. We shook our heads in unison.

"All right," he said. "All of you wait outside except Bassett." The door closed behind us and we could hear muffled voices coming from within the office. After a moment it went quiet, and then suddenly, "thwack." We heard the whip of the cane as it came down, and the sharp biting sound of bamboo against cotton-covered flesh. Then again, "thwack." Again, "thwack." It seemed to take forever to get through all six, especially the last two, and then as suddenly as it had started, the noise stopped. There was the sound of muffled voices and a moment later Bassett came through the door, his face red, but with a huge grin across it. He had survived.

"You next," he said, pointing to me. "He wants you next."

"Shit." I felt my stomach flip and all thoughts of facing the music like a man quickly evaporated. I opened the door and the headmaster was standing behind his desk waiting for me.

"Mr. Hancock," he said, "come on in and make yourself comfortable." I could smell leather again. "I know that you did not have much to do with this incident. You were just there to support Bassett were you not?"

I nodded. "Yes sir."

"Well then, perhaps I should be lenient on you then, shouldn't I?" I nodded again.

"Yes sir." He should be lenient on me. I was only doing it to support my friend and after all, wasn't that what they were trying to teach us at school? I wondered if I might be let off.

"I am being lenient on you, Hancock. That is why you are only getting four instead of six. Now, pick your pleasure," he said, pointing towards the rack on the wall. "You were admiring them only yesterday, were you not?" I walked over to the wall and opened the case. Seven bamboo canes were neatly displayed. They looked like instruments of torture. I tried to remember what others had said about which cane was the worst. I had not expected to be given a choice. Was it the thick one? It looked the meanest. Perhaps the thin cane was the best. It would not weigh as much as the thick one, but then I remembered someone telling me that the thin one was the worst. The cane was so light that it did not seem to hurt at first, but the end whipped around and caught you on your side. I wondered if it would be worse with the long cane rather than the short one. In the end I just grabbed one and handed the flimsy piece of bamboo to the headmaster.

"Ah, good choice," he said. "This one doesn't often draw blood." I couldn't tell if he was joking or not, but nothing seemed very funny. My palms were sweating and my heart felt too large for my chest. Perspiration drowned out the smell of leather.

"Mr. Hancock," he said. "Did you enjoy your evening run last year?" I looked at him in surprise wondering what he was getting at. His eyes were smiling though his face was serious.

"I don't know what you mean, sir."

"Ah, perhaps I should *shine* a little light on the matter," he said. I knew right away what he was getting at. He did know after all. The headmaster knows everything.

"No," he said, as if reading my mind. "We do not know everything. I still do not know who used the saw on my chair. Do you have any ideas?" I shook my head.

"So you are the proud holder of a Victoria Cross then. And what do you plan to do with this achievement once you leave school?" I knew that the game was up.

"N… nothing sir, we just did it for fun," I stammered.

"Don't worry Mr. Hancock, we will let sleeping dogs lie. You were all getting good grades at the time and I thought there was no need to disrupt matters. I heard your dive was splendid." He waved the cane at me. "Well enough of that. Face the wall and bend over." I turned away from him. "And..," he added as an afterthought, "do not forget to say thank-you after each one. This is for your benefit, not mine."

"Yes sir."

I faced the wall and bent over slightly.

"Over more and pull your pants tight. I do not want any baggy pants. Touch your hands to your shins." I bent further and waited. He was enjoying his little game. He knew that there was more punishment in the waiting than there was in the actual caning. A moment later I heard the whoosh of the cane as it whipped down, and then I heard, rather than felt, the whack. There was no pain at first, but suddenly the end of the cane caught me. It whipped around and caught me on my side where there was no protection from my underpants. I jumped with fright since I had not expected it. The pain burned, but I bent down again, grabbed my shins and waited for the second blow. I waited, but nothing came.

"Mr. Hancock, haven't you forgotten something?" I heard the voice from behind me.

"Yes sir, sorry sir, thank you sir," I said. I immediately heard the cane and felt its sting. "Thank you sir," I said. There was another whoosh followed by a sharp sting, and then the last one came. "Thank you sir," I said, and stood up. I was dying to rub my backside.

"Well Mr. Hancock, that wasn't too bad was it?" I nodded, not trusting myself to speak. I wanted out of there, but he kept on talking. "You don't really care about the length of your hair do you?"

"No sir," I said. "I like short hair." I knew what his game was. He was keeping me talking so that I couldn't rub. I knew that he had got me good.

"Tell the next one to come in please," he said. "You choose."

"Yes sir, thank you sir." I left the room in a hurry. "You're next," I said pointing to my friend and bolted down the hallway. All the other students were in class. I headed straight for the changing rooms. Bassett was already there with his pants down. He had been splashing cold water on his backside and was laughing as I ran in.

"How was it?" he asked.

"Not too bad," I replied, dropping my pants to inspect the damage. There were four thick welts across my backside, raised up and bright purple. The marks on my side had all merged into one. There were small flecks of blood on it and it hurt like hell, but I knew that for the next few weeks the marks would serve as a badge of honor. Everyone in the school would want to see them. My stature amongst my fellow students was raised a further notch. I splashed cold water on the marks, and waited for the others to show.

***

High-school graduation marks a turning point in the lives of most young people, but if you were white, male and graduating in the mid 1970's in South Africa, two years of hell stood between you and freedom. There were no patriotic posters urging you to join up, instead an official letter arrived in the mail with your posting. I was summoned to report to the train station in Pietermaritzburg at a specific time to join my fellow recruits for a two year stint as a guest of the South African Defense

Force. Failure to appear would result in dire consequences, or so the posting paper threatened. I was to be stationed in Bloemfontein, a bleak Afrikaans town in the middle of nowhere.

Throughout most of the 1970's and early 1980's, South Africa was fighting a guerrilla war on the western front of its northern border against the Cuban-backed Angolan army, and on the eastern end against Mozambique. All young males were good fodder and there were no exceptions. If you were missing a leg you were assigned a desk job. If you were missing a brain you were promoted to corporal. There would be basic training for three months, specialized training for nine, and for a year thereafter you were posted to the border to defend your country. We were all pawns in a flawed war that claimed thousands of lives on both sides. Not one of us knew why we were there, nor did we know what the ultimate objective was.

The conflict finally ended when all sides ran out of money.

My hair had barely grown long enough to touch my ears when it was my time to report at the train station. My mother offered to drop me off. Neither of us spoke on the way, although I know that there was a lot we both wanted to say to each other. Sons leaving home are a burden mothers accept with quiet dignity, and this was not the first time that she had made the trip to the station. There are four boys in our family. The platform was packed with baby-faced kids saying good-bye and relishing the feeling of being a civilian for the last time. The whistle blew and we boarded the train. As it chugged out of the station I leaned out of the window and looked for my mother's face in the crowd. She had moved away from the tracks and was waving – I could see a tear rolling down her cheek. She looked so young and beautiful. It was the last time I would see her looking well; she died soon after I finished basic training, and with her death I rolled reluctantly into adulthood.

There was no point trying to sleep on the train. Everyone was getting drunk. We swayed and clattered our way through the night singing rugby songs and putting on brave faces. Occasionally someone would lean out of the window and puke into the darkness. For the next three months beer would not be an option and we would be cut off from the rest of the world. Our hours of freedom were waning. When the train pulled into Bloemfontein station we were a sorry lot – unwashed, exhausted and hung-over. The train smelled of stale beer and vomit, and the army was waiting for us.

They say that there are two ranks in the army that wield the power. The corporal and the general. For the next few months we would face the wrath of the corporals, and it started as soon as the train stopped.

"Get your sorry arses off the train right now and *"sak* for fifty!" We quickly found out what *s a k* for fifty meant. Stop everything, drop everything and do fifty push-ups. Do them properly, and do not look up. Especially do not look at the corporal.

"Now *Manne*," the corporals yelled in Afrikaans, "do you see that tree over there? *Fok weg*. Get going." Around the tree we went, and then again and again until the first ones fell and started to puke.

"*Kyk jou maatjies*. Look at your friends," they yelled. "*Hulle slaap nog al*. They have gone to sleep." We were sent to pick them up and drag them back to the train. This running, dragging and puking carried on until late in the day. We weren't even signed in yet, and life had become hell. Bloemfontein is hot, dry and dusty, and the heat was taking its toll. The medic trucks hauled away the first batch, while the rest of us struggled in the dirt. As the shadows lengthened we were marched through the gates of the camp, and with a symbolic clang, they were closed behind us. The camp had ten foot walls around it topped with barbed wire. They were there to keep us in, not the enemy out.

Later that night, after we had been signed in, hosed down with a fire hose and issued army fatigues, we were shown to our tents. Sixteen cots jammed into a small tent was to be our home for the next year. We ate at the mess and returned quickly to the sanctuary of our new home. The lights went out automatically at 9 p.m. I fell into a deep sleep on my first night this far away from home.

At 11 p.m. the sirens went off. The lights came on and the corporals arrived.

"What the fuck do you think you are doing sleeping in your beds?" they screamed. "The beds are for inspection only. You sleep on the floor." And so for the next year we would spend hours making our beds until they were perfect, and then sleep alongside them on the cold gravel floor. The beds had to be absolutely square, and the sheets starched, ironed and folded back a bayonet length; the edge had to be crisp as a blade and a perfect right-angle. We learned how to get them exact. If you marked out precisely where you wanted the edge to be and then rubbed a soft soap along the underside, you could use two books to crease the fabric to a perfect right-angle. By day seven we knew how to bone shoes, clean a rifle, make a bed that would pass the spirit-level test and polish everything in sight to a deep shine. We were tired and homesick, hating the food and the corporals, and only had one year, eleven months and twenty-four days to go. Counting down the days would become a way of life.

Basic training is meant to be hell. The premise is to break

your spirit, and then rebuild you in the mould of a soldier. Initially it was a premise that I cautiously resisted, but I was quickly singled out and worked over until I threw up and begged for mercy. After that, compliance seemed to be an easier course, so comply I did. For our efforts we were paid one Rand and five cents a day, or the equivalent, at that time, of one US dollar, out of which, each month, they deducted three Rands for haircuts, two Rands for mess fees and seventy-five cents for church donations, regardless of your denomination.

The corporals' imaginations knew no limits when it came to creating a miserable environment. They were masters at it. One of their favorite creations was the piss-parade. Once or twice a week we were roused from our makeshift bunks in the middle of the night and made to stand half-naked on the parade ground until we peed. Peeing with an audience is difficult at the best of times, and almost impossible in the middle of winter when there was frost on the ground and three hundred people watching. No one was allowed to leave until the last person had peed. Sometimes we would anticipate the piss-parades and drink gallons of water before going to bed, only to wake up in the middle of the night with a bladder about to blow. So we got up and went, only to return to the warmth of the cot in time to be roused for a pee-parade. There were times when we stood in the cold for hours waiting for the last person to splash before we were all allowed to go back to bed. Some mornings we had just made it back when the 5 a.m. siren blasted, followed by a nasal voice over the loudspeaker.

"*Opstaan manne*. Hands off your cocks and onto your socks. You are in the army now."

In the pre-dawn light, streams of sleepy men made their way to the showers for an ice cold wash. Hot water would only be allowed after basic training. The ablution block was an oblong shed with a row of showers running the length of one side, and a wooden bench along the other side. The bench had holes cut every three feet partitioned by low walls, and served as toilets. No doors. A trough filled with water ran under the bench and every minute a tank at one end would flush, washing everyone's business down the trough into a long-drop well at the end. One morning a friend and myself situated ourselves at the end nearest the tank. We crumpled up newspaper, and just before the flush came, we lighted it and dropped it into the trough. The flush came on cue carrying the burning paper down the trough, roasting bare arses as it went. The corporals found out and we were marched around the parade ground for five hours, but the image of a dozen arses on fire made the time fly by. A strong sense of humor was needed to get through each day.

Guard duty was the responsibility of new recruits, and until a fresh crop arrived after six months, we would stand twice a week. We were issued live ammunition, but told not to use it under any circumstances. The watch system was two hours in the tower, and then two hours back in guard tent trying to get some rest, and so on through the night. It made for an unpleasant night, but the only real danger was getting caught sleeping. It was here that I first perfected the art of sleeping standing up.

We were almost through our basic training when Solly shot himself. He was standing guard in one of the towers and used the field-radio to summon the corporal in charge.

"I am going to kill myself," he announced. The corporal arrived in his jeep with reinforcements and a megaphone.

"Solly you sorry excuse for a soldier, you get down here and let me kick your lily-white arse," the corporal announced with all the tact of a steamroller. "You have disturbed my sleep and you are going to pay for it." Solly said nothing. He was sitting on the floor of the tower staring at the cold cement walls. He slipped the catch off his rifle and put it up to his mouth.

"Solly I am going to kick the shit out of you if you do not come down from there and explain yourself," the corporal continued, his nasal voice sounding more menacing through the megaphone. "You are going to be sorry that you disturbed my sleep you lowest form of life." Solly stood up and was caught in the glare of the corporal's spotlight. His eyes were wide with fear.

"Solly, you stupid shit. If you waste any ammunition you will be sent to detention," the corporal continued, missing the irony of his comment.

Solly blinked back at him, shoved the end of the rifle into his mouth, and without saying a word, blew his head off. The shot rang out across camp. I had been hot-bunking with Solly that night, and heard the shot from the guard tent. It was my bad luck that I was on duty next, and was slated to stand guard in the same tower. The military police dragged both Solly and the corporal away, and sent for me.

"*Meneer* Hancock you are on watch. Get up there and make the place shine." I was given a bucket and mop and told that it would be inspected in the morning. I was terrified. I had never come across death before. I climbed slowly up the ladder into the tower and shone my flashlight around. Someone had already made an attempt at cleaning, but there was still blood and bits of brain splattered on the walls. I shone my light on the roof and saw the hole where the bullet had continued its trajectory after exiting through the top of Solly's head. It was a small neat hole through the tin roof. My stomach lurched, and I threw up on the floor adding to the mess. I felt sick and scared. Life had suddenly become too real for me.

I was not ready to grow up.

A special meeting was called the next day. We were marched out onto the parade ground and left standing in the hot sun. After two hours the colonel in charge of the camp appeared and threatened that there would be real trouble if any of us attempted to do anything as stupid as Solly had done. He shook his finger and glared at us as if we were somehow responsible for Solly's death. It was a tough lesson to learn, but he made it clear; there was no place for pity in the army. No personal counseling sessions. No comforting social workers. It was back to basic training and on with life.

As our three months basic training was drawing to an end, we were told that we would be given a weekend pass to visit our families. We had become skeptical of good news, and sure enough this bit of information was followed by a caveat. We first had a week-long maneuver in the bush, and only those that survived would be granted a pass. Once again we were marched onto the parade ground and made to stand in a long line. At the far end was a pile of tins glinting in the sunlight. From where we stood we could see that all the labels had been removed.

"Right *Manne*," the corporals announced. "That pile over there is your food for a week. One tin, per man, per day. Now, *fokweg*. Get out of here." We bolted across the parade ground with the big guys elbowing their way into the lead. By the time I arrived at the pile all the big tins were gone. A few small tins lay strewn about which I gathered and stuffed into my backpack. I approached a big Afrikaner and asked if he would swap a big tin for one of my small ones.

"You *blerrie* crazy," he said leaving me in no doubt that he was not interested, and I didn't really blame him. Even a big tin would not provide enough sustenance. It was going to be a long, hungry week. We piled into trucks and headed north out of town to a swampy area where our maneuver was to take place. Later that night sitting around the camp fire I discovered that there was indeed someone above looking out for me. The Afrikaner ended up with seven tins of jam. I had oysters.

We slept with our boots on and spent the days crawling through mud and slime. It was hell week and everyone puked and passed out at least once. The full moon had the corporals in a creative state. By the week's end we were certain that they had exhausted all of their draconian fantasies, but they had kept the best for last and sent us on a twenty-mile march through swampy terrain without water bottles. Before we departed, a half dozen sheep were ceremoniously slaughtered and there was a promise of a feast upon our return. The spits were being assembled and the fires were ready to be lit. There was even a rumor of cold beer.

About five miles from the end of our march we were told to start collecting wood for the fire and to carry it back to the camp. I was tired and sick of the army and did not see any point in carrying wood five miles when there was plenty to be found right at the camp. This was just another corporals' sadistic fantasy, and I was not about to fulfill it. I was not alone. A dozen of us felt the same, but I should have remembered high-school and the caning. They had the power and we were the pawns.

As we approached camp, the delicious smell of roasting meat wafted towards us. We had stopped half way and drunk tepid water from a stream, but the effect had long since worn off and I was dying of thirst. I dared to imagine a cold beer in my hand. It had been three months since I last tasted one and memories of the train ride and too much beer had long since faded. I was ready for a good meal and a cold drink. We came out of the trees and into the clearing that had become our camp. In the center, the sheep were hoisted onto huge spits and were roasting slowly over beds of hot coals. Tables had been erected and kegs of beer were placed all around. It looked like an oasis, but first the corporals' had something to say to us.

"Manne, manne, manne." The corporals shook their head in mock disbelief. "You *blerrie* bastards have been here almost three months and not learned a thing." We stared back wondering what they were up to. Just then I was grabbed from behind and dragged off to the side. All the others without wood were singled out.

"Manne, manne. Look at these soldiers. Too *blerrie* smart to carry wood," the corporals sneered. "Your *maatjies* will drink beer tonight and eat until they can't move. You will be crawling through the mud again." My heart sank. I looked around hoping for some wood to miraculously appear in my hands, but it was too late. I was busted and feeling more tired than ever. We were marched off into the woods and crawled and sweated late into the night while the smell of roast lamb and the sound of laughter wafted over to torment us. Life had its lessons, basic training was unfair, they had the power and we were still "the lowest forms of life."

The next day we were issued our first weekend pass. Back in Bloemfontein we were hosed down, issued step-out clothes, given army regulation haircuts, and set free for the first time in three months.

My father loaned me two-hundred Rands and found an old Volkswagen beetle for me to buy. It was turquoise blue, a little rusted and had a sun roof that leaked, but I treasured the car. It was a tight fit, but it could carry five people and all our luggage. The car was my freedom and a way to get back home. In the beginning I would head home on weekend passes, but as time passed it became less important. My primary feathers had started to grow and the tug of life in the big world was pulling me away from the security of family. The tug would grow stronger as my army training progressed, and after my mother died, returning home was traded for visits to a girlfriend.

In the late '70's there was petrol rationing around the world and South Africa was no exception. The government lowered the speed limit and forbade the sale of petrol on Sundays, making it difficult to get back to camp after a weekend at home. I needed more than two full tanks of gas for the four hundred mile trip. Adversity breeds creativity, so I made arrangements to stop half way at a friend's house, and siphon petrol from his mother's car. To make the final part of the trip, I kept a small jerry-can of petrol under the back seat alongside the battery. This was illegal and the police routinely stopped cars to search for containers of petrol, but they never thought to look under the back seat of my car.

A few months after our first weekend pass I had an experience that might have saved my life. Using my ration of cigarettes, I traded a corporal for his parking privileges. New recruits were not allowed to park at the camp, but cigarettes could buy you just about anything, and for a while I was allowed to park my car just outside the camp gates. Occasionally a few of us would slip out on AWOL and head into town for the night. We had civilian clothes stashed in the car and getting past the guard post was easy, but the road to town ran the perimeter of the camp, and there was always a chance of being seen. AWOL was punishable by a minimum of a week in detention, and army detention was much worse than civilian jail.

One evening three of us slipped past the guards and took off for a night on the town. The old Volkswagen had coiled springs and some kind of straw stuffing for the seats. Unfortunately the battery cover had rusted through, and each time we went over a bump one of the springs shorted the battery connections. I drove as fast as I could, but not so fast that I would attract attention, and we had almost made it all the way around the perimeter when I smelled smoke. Puffs wafted up from under the back seat – and then suddenly we were on fire. I slammed on the brakes and we jumped out of the car, grabbed our civilian clothes from the trunk and tried to douse the flames. It only served to fan the fire. Just then another car approached. It pulled up alongside us and the window rolled down to reveal a stern faced colonel.

"And what seems to be the problem here?" he asked. My heart was in my throat. I managed to squeak out a reply that we needed a fire extinguisher. We were suddenly in deep trouble and the flaming back seat was the least of our problems. The colonel produced a small fire extinguisher and passed it to me. I pointed it at the base of the flames and filled my car with foam. The fire subsided immediately. I handed the extinguisher back to the colonel bracing myself for what was sure to come, but he just looked amused and said;

"Have a good day soldiers," and he drove off without another word. He never did realize that we were AWOL. All I could think of was what would have happened if we had a jerry-can of petrol under the seat.

The first year passed soon enough, and the sign pinned to the back of the tent door read "365 days to go." Our status was raised from being "the lowest forms of life," to "*ou Manne* –

old men." We now had some of the power. I graduated at the top of my platoon. The top twenty were selected to be part of the elite tank division. The rest were sent off to train in armored cars. We would spend our days out on the shooting range lobbing shells at targets three miles away. It was a pathetic waste of money. We gunned down cactuses, and any bird that had the temerity to fly close by was instantly blown out of the sky. In fact, anything that moved that wasn't human, was mowed down. We were kids and we were bored. The armored car divisions were sent up to the border to fight the war against the Angolans'. Because the terrain was thick bush and no place for a tank, and so we were left behind to sit in the sun, polish the bright-work and change the oil.

The boredom was somewhat relieved when a fresh crop of recruits pulled into camp. We were now in a position to make their lives miserable. We had been well taught and it was time to pass along some of our hard-won knowledge. We conducted inspections every morning, running our fingers along the tops of the tin cupboards, knowing full well that we would find dust – moments after you clean it, dust settles again.

"So *manne*," we would say. "You are a bunch of pigs. You live like savages." The new recruits would tremble and stare at the ground while we did our best to suppress a laugh. It was all so ridiculous.

"So *manne*, you know what pigs can do if you train them properly?" They knew the answer. They had heard it many times before.

"Pigs can fly," we would say with dramatic emphasis. "So pigs, it's time to fly."

The recruits would hand us their steel helmets and climb up on top of their cupboards. We would remove the soft inner-lining and pass the helmets back. They stood on the cupboards, helmets on, strap unbuckled, arms outstretched and when we said fly, they jumped. For a moment they were airborne and then they landed softly on the gravel floor. A split second later five pounds of steel helmet landed on their heads. It crashed down knocking most of them to the ground, and in some cases knocking some of them completely out. It was great fun now that we were the instigators, but I'm afraid generations of soft-headed South Africans were created this way.

Each morning we marked off another day on the calendar and pinned the result on the back of the tent door. Slowly, the magic number decreased. When it read "100 days to go," we had a huge party and then went back to polishing our tanks. Bloemfontein had become home, and despite the dust and the heat it wasn't such a bad place. We fought boredom and watched the days go slowly by. With just forty days remaining, I was called into the colonel's office.

"*Meneer* Hancock," he said, "you are a sailor, are you not?" he commanded rather than asked.

"*Ya colonel*," I replied.

"Well then *meneer*, you take a train to Cape Town and you represent this camp in the Defense Force Championships. They are racing a regatta and you make *blerrie* sure that you win."

"*Ya colonel*," I replied, and took off to the purser's office to get a cash advance. He handed me a train ticket and the following day I reported at Simonstown Naval Base. The navy appeared to be much a more relaxed place than the army. I was given a private room and a boat to sail. We would race in False Bay in clunky old dinghies. The wind howled down the back of Table Mountain, and the regatta turned into a survival of the fittest. I won all six races beating the army's, navy's and airforce's best, and was awarded a trophy that became my ticket out of the army.

I took the train back to Bloemfontein while the trophy was flown on an airforce jet; it went first class, while I was still going third. The train rumbled across the dusty Karoo, and as I watched the scenery pass, I thought about how much I had changed since my first train ride to camp almost two years earlier. I had grown up. The army was not a place I would have chosen to spend two years of my life, but the training and discipline had been good. I had the rest of my life to look forward to. Sailing for a living was a lifestyle that I was slowly daring to dream about. As night fell, the landscape faded through hues of orange and red, until finally turning black. I made up my mind to leave South Africa and look for work in Europe. In less than a month I would be free to do anything I wanted, and thought of traveling the world looking for sailing regattas to participate in.

When I arrived back at camp, I was met at the gate by the military police. They seemed friendly, but my first instinct was suspicion. I tried to think what I might have done wrong.

"This way *meneer* Hancock," they commanded, and led me to my tent. "Pack your stuff and make sure that your rifle is clean. We will be back in twenty minutes to get you. And...," they added as an afterthought, "stay in your step-out clothes." In a moment they were gone and I sat on my bunk wondering what was going on. The tents were deserted. In the distance I could hear a corporal drilling new recruits on the parade ground. His voice carried through the empty camp. I sighted down the barrel of my rifle, saw that it was clean and tossed it onto my bunk. I emptied out my tin cupboard and stuffed everything into a duffel bag wondering where they were transferring me to. The Military Police returned and took me to the quartermaster's tent where he took the rifle and checked the contents of the duffel. They then marched me out to the parade ground.

As we came around the corner I saw the whole regiment assembled. A stage had been set up and the top company brass were all seated. On a table front and center, stood the trophy. I could not believe my eyes. After two years of being "that *fooking* Englishman," I was being recognized with a parade. The band struck up, and when they stopped the colonel stood and made a short speech in Afrikaans. He then summoned me to the stage and presented me with the trophy.

"*Meneer* Hancock," he said, "you have brought honor to our camp." I stared at him, not quite believing what was happening, and shook his hand. "You are *fooking* lucky you did not lose the *blerrie* races. I have something for you." He handed me an envelope. "You are finished with the army. Those are your get-out papers. You may go home now." I stared at the envelope. "You may go now," he said. I saluted him, turned around, and as I walked off the parade ground I heard him call out behind me. "Jus' make sure you make something of your life hey. It's too short to waste." The amount of compassion in his voice took me by surprise. The colonel was not known for sentiment. The parade ground looked different as I walked away. It was no longer a place to fear. I stepped onto the grassy verge and ran down the slope towards my tent. After more than seven hundred days I was finally a free man.

I turned in my step-out clothes, signed out at the guard hut and walked to my car with the few possessions I had managed to accumulate over the past two years. It was not much to show for all the grime, sweat and tears, but I had arrived there as a boy and was leaving a man.

I had hair on my chin to prove it.

SAILING THE WIND OF FREEDOM

"All men dream: but not equally. Those who dream by night in the dusty recesses of their mind wake in the day to find that it was vanity: but the dreamers of the day are dangerous men, for they may act upon their dreams with open eyes, to make it possible."

T.E.Lawrence

Race day dawned bright, but my mood was dark, the effect of too much beer, too little sleep, and the news Skip had given me as we tossed the dock lines ashore.

"We've hardly sailed this boat," he said, the crew have no clue what they are doing, they don't speak English, we've run out of money, and you're on the other watch." Skip delivered the news in the same cavalier manner to which I had become accustomed, but this time there was six thousand miles of open ocean ahead of us, and my heart sunk. It sunk a notch further when I noticed one of the crew spinning the winch to see which way it turned before he wound the line around. They really had no clue. No idea what they were getting themselves into and no sense of the difficulties that lay ahead.

I envied them.

I had arrived at Ocean Village in Southampton the day before, jet-lagged and tired. The previous twenty-four hours had been a mad scramble to get ready for the race. Skip Novak, my good friend with whom I had sailed many thousands of miles, called on Thursday morning and asked me to join his crew for the first leg of the 1989 Whitbread Round the World race.

"It should be fun," he said, "the guys are great, and the boat seems fast. You need to be here by tomorrow night," and so it was Friday afternoon when I dragged my bags out of the taxi, and made my way towards the marina where the Whitbread fleet had assembled. It was a damp English afternoon — the air smelled of rain, despite a bright forecast. I left my luggage with security, and was about to order a beer when I noticed a crowd gathered along the breakwall. I wondered what they were watching. As I elbowed my way to the front of the pack, I could see a sleek, low-slung yacht short-tacking up Southampton water. The Soviet flag snapped briskly in the wind as the crew maneuvered the yacht through a throng of spectator boats. White letters along the waterline announced that *Fazisi* was her name. My ride had arrived and all seemed well, but my sailing and communicating skills would soon be tested. This was not going to be an ordinary sail across the pond. I was about to become the "other" non-Soviet on board the Soviet Union's first-ever Whitbread entry.

I fought my way through the crowd on the dock, and scrambled on board while Skip did a quick introduction, and then he departed for more pressing issues. I was left to my own resources. I nodded at the crew who were still not quite sure who I was or why I was there, and made my way down below. The crisp red and white paint job of the exterior contrasted sharply with the dull gray of the rough interior. The low-slung freeboard that gave the yacht its unique *"watch out here I come*

Low freeboard on *Fazisi* made it a
wet ride when the wind was up.

look," did nothing for the inside, and there was only a small crawl space between the rear of the yacht, and the front. Headroom was minimal and reduced to nonexistent when the sails were dragged through the cabin.

The comforts of home were going to be few.

I found my bunk jammed high in the aft end, nestled amongst the salamis. The cook was busy stowing food, and had a variety of sausages strung from the overhead. Later in the trip when the tropical sun beat down on the deck, the salamis would drip oil all over the sails, but for now they added a certain charm. I stowed my gear, wondered for a moment if I had lost my mind, and then headed for the beer tent. There was nothing a few pints of warm pub beer couldn't cure.

The Whitbread would start in less than eighteen hours.

As the gun fired I was filled with a sense of elation and dread. It was an uneasy combination, but the dread was soon washed away by the excitement of the crowds and the unbridled enthusiasm of the crew. They were just happy to be anywhere but the Soviet Union. The recent odds handed down by the London bookies did nothing to deter their excitement. They had been betting 100:1 that the Soviet crew would not make the start, let alone the finish, and we had just proved them wrong. If victory comes in small packages, we had opened our first, and there would be many more. There would also be unwelcome packages, but for now we were Russians headed for fame and fortune, just as Jason and his Argonauts had done, two centuries before.

* * *

My sailing career was already well established. It had begun a decade prior to joining *Fazisi*. The only way I could get out of South Africa on no budget was to sail out, and so I did. I left in 1979 aboard a South African yacht called *Dabulamanzi*, and half way across the South Atlantic decided that I wasn't going back. Instead, I was thinking ahead and planning to sail around the world. It was a perfect way to avoid the realities and responsibilities of life, and a great way to meet girls. I was hooked. In 1979 I did my first long race from England to Australia, with a stopover in South Africa. It was called the Parmelia Race, and was a celebration of the discovery of Western Australia. By the time we arrived in Fremantle I was already thinking about doing the longer and tougher Whitbread Round the World Race. I knew it would be difficult, but it was the ultimate yacht race, and the appeal was huge. I planned to be in England in 1981 to join a boat.

I joined the American yacht *Alaska Eagle* as their sailmaker. The yacht was quickly dubbed *Alaska Beagle* by the rest of the fleet. They started off calling the boat *Arctic Crow* in jest, but when it became obvious that the boat was a dog, "Beagle" seemed more appropriate. It had been modified for the Whitbread, but the changes only slowed it down, and we thrashed our way around the world trailing the fleet and dealing with crew dissention and desertion. The cook was an alcoholic with a bad attitude, and we ended up placing one of our crew under psychiatric observation in New Zealand. We might all have benefited from analysis, but instead we left for Cape Horn – down one guy – and managed to slog on to the finish in England, arriving on a cold, gray UK afternoon. No money and no prospects, but one hell of an adventure under our belts, and only three years until the next one.

They say that it helps if your IQ is less than the length of your boat, especially if you plan to do a second Whitbread, and while I am sure there is some merit to that statement, I simply wanted to do the race again to do it properly. This time on a good boat, with good people, well-funded. The race committee cleverly spaces Whitbread races four years apart – it takes that long to forget the misery – and I had long since forgotten the cold and wet wild sailing, and happily joined the crew of *Drum* in 1985 for another go-around. The boat became famous before the race even started when the keel broke off and the yacht capsized. The crew, I was not amongst them, were airlifted off, and *Drum* was eventually salvaged and refitted and readied for the race. *Drum* also turned out to be slow, but the guys were great, and the campaign had an added flair because of the celebrity status of the owners. Simon Le Bon, the lead-singer in the rock group Duran Duran, along with the two managers of the band, footed the bill and sailed on board. We were greeted at each stopover by screaming Duran Duran fans, but as one of the crew dryly pointed out; "it would have been better if the boat was sponsored by Julio Iglesias, at least his fans were of legal age."

By the time 1989 rolled around I had hung up my Whitbread boots and was living a quiet life on Cape Cod, dabbling in real estate and raising a daughter. I had not given any thought to another race around the world until Skip's fateful phone call that now had me gesturing madly at the Soviet crew in an attempt to get them to raise the spinnaker. The foredeck hand was awash in the scuppers with the spinnaker in one hand, and the spinnaker sheet in the other, while the rest of the crew were waving at the spectators oblivious of his plight. It was going to be a long, hairy passage to South America.

* * *

The Solent is a narrow strip of water between the Isle of Wight and the British mainland. On the day of the start it was churned from its usual dull gray, into a foamy white, whipped up by a mass of spectator boats and twenty-three world-class yachts headed for Uruguay, six thousand miles away. Our low freeboard had us plowing through the slop and chop with the boat half submerged. By the grace of god and dash of luck we had our sails up, and the boat pointing in the right direction when the start gun fired. *Steinlager*, the powerful ketch from New Zealand was on our starboard hip, and the rest of the fleet were tucked well to leeward, and behind. It was the best showing *Fazisi* would ever have. I trimmed the genoa on the port side and fought the urge to throw up. The realization of spending a month at sea on an untried yacht with an inexperienced crew had just hit me, and the thought sunk to the bottom of my gut like an unexploded depth charge. It would take many days of fair sailing in the trade winds before the feeling went away.

My first problem was reprogramming my mental rolodex, and to try and get a grip on the names of the crew. Common names are hard to remember. Uncommon names delivered in a strange tongue are nearly impossible. Igor and Viktor were easy, but Nodari and Gennadi were not. "Hey you!" didn't cut it because of the language barrier, so I ended up remembering those with nicknames first, and gradually figured out the rest with time. The "Elephant," and the "Crocodile" were extroverts and became my immediate conduit to the rest of the crew. They understood more English than they let on, and had a good grip on sign language and universal gestures. Juki didn't speak a word of English, but was one of those rare individuals who is able to communicate perfectly with hand gestures and rich body language. Most of the others were going to take some time, but as we plowed out of the Solent and into the English Channel, I had other matters on my mind.

There was the immediate problem of sailing the boat. Maxi yachts are heavy and cumbersome, and dangerous to the inexperienced. The after-guys are half-inch wire and strung bar-tight under load. A mistake could cost you a finger, or worse, your life. To my surprise and delight *Fazisi* seemed to sail herself just fine. She was much lighter than her counterparts, and slipped through the water leaving barely a trace. A few bubbles in the wake were the only sign that we had passed by. The crew had (eventually) hoisted the spinnaker, and we carried the sail well

Deck work on *Fazisi* was made more interesting by the language barrier.

into the Bay of Biscay, gybing occasionally, and changing to a lighter sail as the wind fluctuated. Skip plotted a route that would take us to the northwest corner of Spain, and then out into the Atlantic, before crossing the equator and closing land again at the bulge of Brazil. It was a simple strategy. We were taking it a day at a time, adjusting our plan to suit the mood of the crew and the speed of the boat. We had no illusions about winning – finishing the leg in one piece would be a victory, and who knew what lay beyond the finish line? The soviets had run out of money, and Uruguay was hardly a Mecca for sponsorship dollars.

***

The *Fazisi* project was the brainchild of Vladislav Murnikov, a civil engineer from Moscow. He had read a copy of *Sail* magazine and seen photos of an earlier Whitbread fleet. Undeterred by the practical impossibilities the Soviet Union presented, he decided to launch a soviet effort. The country was flirting with change, and he reasoned that a high-profile world-wide sailboat race would be the ideal vehicle for promoting soviet industry. The chairman of Fazis Company, an import-export business based in the soviet province of Georgia, agreed, and the Golden Fleece Syndicate was born. They would build a high-tech yacht and go up against the world's top ocean racers, proving that there is no substitute for pure optimism, focused energy, and most important of all, blind faith.

The theme of their campaign would be the Greek legend of Jason and the Argonauts, and their search for the Golden Fleece. The legend has it that Jason and his Argonauts landed in Colchis (now Soviet Georgia), where Jason surreptitiously rowed up the river Phasis and recovered the fleece hanging from a tree where it was being guarded by a serpent. Metaphorically speaking, he who finds the fleece, finds success, and for the soviets their "golden fleece" was to be western sponsorship. What most people forget, and the soviets chose not to remember, was that recovering the fleece was the easy part. The return trip was fraught with danger and problems. If anyone had dared to contemplate the tale in its entirety, they might have seen a warning of events to come.

The soviets had appointed a co-skipper for the race. A bull of a man with a moon face and shallow chin, Alexei Gryshenko epitomized my stereotyped impression of all Russians. He was, in fact, from the Ukraine, but looked and played the part of "party leader." He shook my hand when I was introduced, grunted something I didn't understand and barely acknowledged my presence on board for the duration of the trip. He sat apart and aloof from his crew, occasionally beckoning one of them to join him, like a lord and his serfs. I had no interest in crew politics and ignored Alexei for the most part – a choice I would later regret.

On our fourth day at sea I came on deck to see Alexei sitting alone at the back of the boat. He nodded to me and grunted a greeting. I wondered about his frame of mind. Building the boat in Georgia and finishing it in England on an insecure financial footing had taken a toll on all the crew, none more so than the co-skipper, but I figured a few days of sunshine and warm weather would sort him out. On the horizon ahead we could see two Whitbread boats, and behind us, a third. We were doing OK. The weather had been good and crew morale remained intact. I found time to figure out the deck lines and halyards, and learn some of their Russian names. I had one of the crew translate "I need food," and "I need sleep," and I used the two phrases as often as I could. It seemed to amuse the guys, and got me what I wanted.

By the end of the first week I thought I had their names all figured, until one day I noticed a new crewmember joining us for dinner. I had not seen him on board before, and wondered how I could have missed him. He also showed up on my watch and I asked Skip if he had traded a guy from his watch, and taken Juri, who was missing from mine. Skip laughed and pointed to a razor and mirror in the bucket on the aft deck. Juri had shaved his expansive beard, and shed thirty pounds in the process. They were not making it easy for me.

They were not making it easy for Alexei either, or perhaps he was not making it easy for himself. He was slowly becoming more remote, and spent much of his time on deck alone with his thoughts. I had given up greeting him, and he had given up grunting at me. I wish I had made more of an effort, but it's too late now.

As we approached the equator, routine on board settled into a rhythm. We were languishing in last place, lagging way behind the second-last boat. The crew had slowly sorted themselves into a loose-knit team. Without the language I missed the conflicts and struggles that were being waged among them.

To me, all appeared fine.

The team had been drawn from across the Soviet Union, and despite having grown up under a single soviet decree, their cultural backgrounds were diverse and varied. The only constant was their lack of personal motivation and intuitive sailing ability. They waited to be told what to do, rather than getting on with the job at hand. Sailing a powerful maxi-boat like *Fazisi* required a lot of forethought and instinctive actions, and Skip and I were doing our best to hammer some western themes into them. We appeared to be failing. It's hard to undo a lifetime of different thinking.

I did notice that the cook was being used as a political football, but then that's the lot of sailing cooks the world over. Good ones give what they get, and Alexei (the cook), was doing just fine. He was dishing up a combination of freeze-dried fare and soviet specialties, handed out with a dash of abuse. His was a difficult job. With no standing headroom below, he was forced to sit down to cook. The galley was built under the cockpit, and was hot, cramped and stuffy, but despite the squalid conditions, the food was passable.

Alexei had been recruited at the last minute and doubled as ship's doctor. His most recent assignment had been as a bush doctor in Afghanistan, and his bedside manner reflected it. It was nonexistent. So too were his cooking skills. His lack of culinary expertise was exaggerated by the fact that the food was freeze-dried. He had never seen western food, let alone western freeze-dried food, and he couldn't tell the difference between powdered egg, powdered milk and mashed potatoes. Some mornings we would get mashed potatoes for breakfast, and powdered eggs in our coffee. Other mornings it was simply a plate full of chopped garlic and onions. The soviets loved the garlic and onions.

"Good for the blood," they said. Out of self-defense, Skip and I followed their lead, and soon discovered their secret. Half way though my first onion my throat turned numb and stayed that way for a week. It made the cook's cooking palatable.

I was learning quickly.

We charged through the doldrum belt seeking out rain squalls with our radar, and hooking onto the windy side of each one. The brain trusts of the bulk of the fleet were ensconced in front of computer screens, crunching numbers and analyzing weather data, but we had no sophisticated weather receiving equipment, and relied on the old fashioned method – we looked out the window and dealt with what we had been given. This soon had us in fifth place, snapping at the heels of the leaders.

In our enthusiasm we called our race headquarters in Moscow.

"Great news," we said. "We are no longer in last place. We're now in fifth."

"Ah, good, good," they replied. "Good news, but it makes

Crossing the equator with the soviets.

no difference really. The Russian press has been reporting you in first place since the start anyway!" It was my first lesson in the power of propaganda. We were guaranteed to win no matter how badly we finished – as long as we finished. Skip and I were quickly learning the Russian way, but they were not mastering ours. They were still waiting to be told what to do.

"Think for yourselves," Skip would yell at them. "For Christ sake, don't wait to be told what to do. If someone falls overboard are you just going to sit there and wait for us to tell you that they are drowning?" The guys sat with blank expressions and waited for Skip to finish. It was not in their makeup. They looked sheepish, but said nothing. For the next few days Anatoly didn't come on deck. He spent his time at the nav station with a pencil and reams of graph paper working feverishly on some plan he was concocting. He reappeared with a triumphant look on his face.

"I have solution," he said, and handed each crew member a sheet of graph paper. In neat script he noted the time of day for the following week in half-hour increments. Across the top of the page were a list of duties that might need doing, such as grinding the winches, helming the boat, or trimming sails. Penciled in, were the duties of each member of each watch for each half hour of each day. At the time change they were to consult their sheet and move to their next duty, no questions, no comments. It was a beautiful piece of social dissertation, but doomed for failure. It never took into account what sail combination might be up, and fell flat when Juki realized that he was relegated to winding winches for most of the rest of the trip, while Anatoly would be doing eighty percent of the driving. I buried my hands in my face and wondered why the Western World ever feared the Soviet Union.

For the most part, I was keeping to myself. It was difficult to communicate with the crew, and with Skip on the other watch, there was no one else to talk to. We passed each other in the companionway at watch change. It was tiring work, and I usually headed for my bunk as soon as I could. Sometimes Skip would climb aft and sit on the sails and we would discuss the crew, the boat and the trip in general, but mostly we would talk about life on land, passages past, and future projects. Skip had a lot invested in the success of the soviet effort. His name was up there in the headlines, and if the project faltered, he would falter right along with it. I did not have much to lose. I was along for the ride. I knew that there would be some good stories to tell after the finish, and time at sea was always time well spent.

I was enjoying life.

On day twenty we converged with land and passed close by Recife. A couple of days later we saw the loom of Bahia and I remembered my visit there a decade earlier. It had been quick and memorable. A day later we blasted through a maze of oil rigs off Cabo Frio, and saw the loom of Rio far to the west. The lights faded into the dawn, but the sight of land stirred memories, and we were starting to crave civilization. Three weeks of Alexei's cooking had us hungry for home cooked meals.

We were still holding onto fifth place, but once the trade winds settled in, the powerful ketch, *The Card* rolled us. She sailed abeam for a few hours, and then pulled ahead with her mizzen gear adding horsepower. It was time to get to Uruguay. The crew were restless and Skip and I were ready for a break. Our only concern was the well-being of the co-skipper. Alexei had become more distant and less communicative. I wondered if he was pleased or disappointed with our placing, but never thought to ask.

When we closed land again, it was the flat, featureless coast of Uruguay. White sandy beaches sparkled in the hot summer sunshine, and waves washed up on the shore. The sight was a feast for the crew whose last look at land had been the coast of France. Much had happened since. They had molded themselves into a team of sorts, and their English was improving. As they told me later when I asked why they were so keen to speak English, they said, "In port we need meet girl, we need speak English." It was the old universal motivator that had them studying the language, and what's more, it appeared to do the trick.

Many hard-won miles and a respectable placing should have left a deep sense of satisfaction among the crew, but I noticed some tension building as we neared the finish line and wondered what the stopover would bring. The rhythm of life at sea abruptly gives way to the hard reality of life on land, and I knew the stop in Punta del Este was not going to be easy. It was a clear, moonlit night when we powered across the finish line, four weeks to the day after leaving Southampton. The guys had come to accept me as part of their crew and acknowledged my input. It was all I could ask from them. My contract was up, and with a sea bag full of memories and some new friendships to count on, I headed to the airport.

Ocean Village and the start of the race seemed a lifetime away.

***

I flew to England and took a train down to Cornwall for a family vacation. The lush English countryside was all the more pleasant after a month at sea, and the village of St. Just-in-Roseland was a picture perfect place to ponder my future, far removed from the hustle and bustle of Punta del Este. Each morning I would walk along the shore and look out to sea. Cold fall weather had settled over the British Isles. Out in the Atlantic Ocean great sheets of rain gathered to drift slowly towards land, and wet weather drenched the coast. The Whitbread fleet had escaped to the trade winds in the nick of time. I thought about my trip aboard *Fazisi*. It had been a unique adventure and an interesting learning experience. I hoped that the crew would find their golden fleece. I hoped Vladislav's dream would come true.

In the words of Walt Whitman — "I dream … in my dream all the dreams of the other dreamers, and I become the other dreamers." The world needs more people like Vladislav. The dreamers of the day are indeed the dangerous ones.

On my way back from the walk, I stopped in a café for tea. It was the only place in the area that sold newspapers and I bought a copy of the Daily Telegraph in the hope of gathering news about the remaining yachts finishing in Uruguay. I flipped to the sports section, and glanced at the bottom of the page. The headline sent me reeling.

RUSSIAN SKIPPER FOUND HANGED FROM TREE.

I did not have to read further. I knew all along that something tragic was going to happen. I had developed a feeling deep in my gut. I gazed at the peaceful Cornish countryside and wondered about his wife and one-month-old baby left behind in the Soviet Union. I wondered about the crew. I wondered about Alexei. What had compelled him to take his life? What was he thinking as he sat alone on deck staring at the water rushing past the hull? What was he thinking as he sat alone on the branch seconds before taking his life? "What if…?" — the saddest sentence in the English language. Or is "if only" sadder? If only I had taken the time to sit and talk to him. If only Vladislav was not a dreamer. If only.

The air felt much colder as I left the café. It smelled different. More rain was imminent. I walked back to the shore and sat in a field looking out to sea. I sat for a long time, thinking about what I might have said or done different, and eventually fell asleep. When the rain came it woke me. I blinked my eyes open, and standing all around were a dozen cows. They were gazing down at me with their soft, gentle, brown eyes. It's OK they told me. It's OK. The world spins on its axis, but because of the tilt, some people fall off. Perhaps they go to a better place. Perhaps they were too good for this world.

Whatever the case, Alexei had sailed his last passage.

A SERENDIPITOUS PISS

"In time, and with water,
everything changes."
Leonardo da Vinci

Bahia is a city that throbs. There is no better word to describe it. The tropical heat and clinging humidity mingles with the sweat of lovers who stroll the beaches and boardwalks by night, while samba music filters through palm fronds. Rhythm hangs heavy in the air. Occasionally a cool breeze wafts down from the mountains and blows the laden air out to sea, but mostly it just sits there like a pregnant woman; uncomfortable and hot. The city can be intoxicating. The stagnant warmth stirs feelings long-since latent. It awakens primal urges. Bahia is a city with a soul, and when you step ashore it wraps around you, hugs you tight, and reminds you that you will never be the same. When you leave, a small piece of Brazil leaves with you.

I first visited Bahia in the summer of 1979 while on a passage between Uruguay and the Caribbean. I was twenty-one years old, sailing the world and searching for an identity to call my own. The loom of land beckoned from a hundred miles out, and by dawn we had altered course for a visit. The passage had been

Dolphins escorted us up the coast of Brazil. They're my good luck charm.

a tedious slog into a strong northerly headwind, and the boat and crew were getting tired of the grind. We were ready for a few days of fresh food and a firm footing.

It was a little after noon when we passed the breakwater

and sailed smack into the embrace of tepid air. The water changed from a deep blue to milky brown, and smelled of run-off and rotting vegetation. A small armada of local crafts jostled for position, hawking produce and trinkets. We were escorted to an anchorage. As the hook went down I stood on the bow drinking in the sights and sounds of land, and longing for some time ashore. My hormones were cranking and a night on the town was on my mind. The back alleys with their forbidden fruit beckoned. I decided I would slip ashore later that night to see what I could find. I had no idea that my life would be almost ended in the next few hours.

There is a huge public elevator that transports passengers from the dock area, up to the main part of the city. You leave the quiet squalor of the docks to land among the hum and bustle of the old city where street vendors air their wares, and buskers tap out a rhythm that echoes on the walls of the old buildings. Open fires burn on the sidewalks with strange pieces of meat grilling over low coals. The fatty morsels are sold to the passing trade. Thick smoke hangs in the air, blurring the lights and adding to the heady feeling of being in a foreign country.

As I headed up the elevator, I felt like a stranger in a strange land. I enjoyed the feeling and the perspective it gave me, and studied the faces of those around me wondering what they had made of their lives. I recalled Ibsen who once said — *there is always a certain risk in being alive, and if you are more alive, there is more risk.* For me, feeling "more alive" is an addiction without antidote, but Ibsen was right, it can be risky business. I studied the character lines on the face of an old woman standing next to me. She clutched her basket closer to her breast and eyed me sharply. I smiled at her, and saw her shoulders drop and the furrows on her brow smooth in response. She shuffled her feet and smiled back. If wrinkles were anything to go by, she had lived a full life. A wide-eyed kid with a snotty nose stared at me without blinking. His round face glowed with innocence – yet to be marked by the passage of time. Behind him a businessman gazed at the floor with a bored look. I felt bad for him. He looked uncomfortable in his conforming dress. Patches of sweat marked his suit. The elevator was hot and stuffy and I was glad when it came to a halt. I slipped the kid a few cruzeros and headed into town.

It was still early, and the streets were empty, but as I ambled towards the central part of town I felt the city come alive. Street vendors hawked everything from live parrots to dead rabbits. The street urchins were out in force and I patted the back of my trousers, checking for my wallet.

"Mister, Mister," they yelled at me. "Mister, Mister come and look." I smiled and kept moving while shoving my wallet deeper into my pocket. A group of kids had set up drums and were banging away to some inner beat, their faces transfixed by the rhythm. A crowd had gathered to watch, and occasionally someone would toss a few cruzeros into a hat. The takings were meager, but the evening was still young. As the sun disappeared behind the mountains an occasional rumble of thunder rolled down from the hills. The light on the old buildings changed, throwing shadows across the square and softening the hard edges. Looking out across the bay I could make out our boat at anchor. A shadowed figure on the foredeck was rigging the anchor light. A dim glow emanated from the portholes reflecting on the still water. I thought of the crew sitting below preparing dinner, slugging back beer and sipping red wine. None of them had wanted to come ashore, opting instead for an early night. I was glad to be alone. I felt like an anonymous stranger, a refugee from another world, privileged to catch a glimpse of life in Brazil.

I kept moving, drinking in the sights and smells, feeling the beat of the city resonate in my gut. Soon I was beyond the main square, into the back alleys where the streets narrowed. While Bahia can hardly be described as a city for tourists, the central area accommodates commercialism much better than the alleys. Some of them were unlit, most of them filthy. I was searching for a hole-in-the-wall restaurant, and knew that the farther from the square I went, the better my chances would be of finding something in my price range. I glanced into a couple of restaurants but they were deserted. Brazilians seldom venture out for dinner before midnight. Bored waiters hung around the doorways, chatting amongst themselves and commenting loudly on the passing crowd. Macho men swaggered by, chain-smoking unfiltered cigarettes, most sporting a forest of coarse hair protruding from their open vests. Girls in tight jeans swung their hips and pouted their over-painted lips in response. Not a soul paid me heed. Not that I blended in very well with my blond hair and blue eyes, but no one seemed to give a damn.

I stopped at the old cathedral, went in and sat in the cool darkness watching people come and go. The confessional was doing a brisk trade. An old priest came in through a side door and nodded to me as he passed. I smiled at him and watched as he lit candles around the alter. I could hear the kids banging their drums, though the sound was muffled by thick walls. A young girl sat next to me and showed me her rosary, but she left when she realized that I could not speak Portuguese. I wished I could. There was something sublime about the old building. The stained-glass windows had been there for hundreds of years, and the pews had provided sanctuary for many weary travelers.

I was only the latest, not the last. It was an oasis away from the bustle of life on the streets, and open to all. I enjoyed the quiet for a moment, but the pull of the street was strong, and I was dying for a cold beer, so I headed back out the way I had come in.

Out into the heat and humidity.

By 10:30 p.m. the bars had started to fill, and elderly couples were nursing drinks in some of the more well-lit restaurants. I found a narrow bar down a side street, and slid in for some refreshment. Cigarette smoke hung thick, engulfing the piles of sausages and salamis that were strung from the ceiling. I wondered how much of the curing process took place in cigarette smoke. The bartender nodded as I took a seat by the door, and an icy Brahma Choppe appeared in front of me. My eyes smarted from the acrid air, while my throat gratefully accepted the beer. At first glance the place looked to be full of men only, but then I noticed a couple of women sitting by themselves in the corner. Hookers? Perhaps, not that it mattered. They were keeping to themselves and seemed to be surveying the scene as much as I was. The crowd was relaxed and laughter pealed out from all corners. The beer was good, but I wasn't sure how much of the smoke I could take, so I signaled the bartender. Did he have any suggestions for a good restaurant? "Something where the locals eat, preferably locals that do not have much money yet enjoyed a decent meal every now and then."

"*Sim*," he beamed, "*Sim, Sim. Tudo bem.*" Yes, no problem. He had a brother that liked to eat. He rubbed his stomach vigorously to emphasize the point. "I like to drink," he said, clutching a bottle for effect. Two blocks away. A good meat place. It sounded fine, so I drained my beer and headed out into the warm night. The throbbing had started. Bahia was coming alive.

"Psst Mister, you wanna watch." A kid no older than ten tugged at my sleeve. "Wanna watch?"

"Watch what?" I asked.

"No Mister, I gots watches, all kinds." He pulled his jacket back to reveal an array of the latest sports watches tacked to the lining of his grubby coat.

"No thanks," I said and kept walking, checking my back pocket for its familiar lump.

"Mister, Mister you wan' smoke?" he persisted, "I gots American cigarettes."

"Nope, I don't smoke," I replied. I removed my wallet from my back pocket and slid it inside the front of my trousers.

"I see you later Mister. I bring nice girl." With that he disappeared into the crowd.

A few blocks away I came across the restaurant. On display in the front window was a small lamb splayed on a spit, roasting slowly over hot coals. The juices dripped onto the ashes, quickly igniting and sending puffs of pungent smoke into the air. The restaurant was crowded. Smoke filled the place, although this time it was the greasy kind from the fire pit at the back of the restaurant. The waiter had looked askance when I asked for a table for one, but said nothing and showed me to a small setting alongside the bathroom. I guessed that eating alone was considered bad form.

The entire back of the restaurant was an open fire pit with logs piled on top of each other. They were held up against the wall by a sturdy wire grate. The chef threw new logs over the top of the grate to feed to fire, and extracted red hot coals from the bottom to cook the meat. I watched him use a stubby rake to spread the embers over the length of the grill, piling coals where heat was needed. Thick slabs of meat sizzled on the racks. Off to the side an assistant attacked a carcass with a cleaver. A loud "thwack" resulted in a portion of the carcass being dismembered in a splattering of blood and bone chips. It was shoved aside for grilling. On the table in front of the chef lay a pile of intestines ready for separating and grilling. Nothing would be wasted. I noticed that some of the beef still had the skin attached. It was separated from the meat by a layer of yellow glutinous fat. Special cuts for those willing to pay extra. These pieces are slow grilled over low coals, skin side down. The heat permeates through the hide, heating the fat which sizzles and splutters, slowly roasting the meat. I ordered a slab with a green salad side and fried potatoes dripping in oil – guaranteed to clog the arteries, but so good. A jug of house red came with the meal. It was served in a terra-cotta carafe and poured into a heavy glass.

I separated the meat from the fat, and set the hide aside. It was tender and tasty and more than made up for the discomfort of sharing my plate with the singed waste. The beef melted in my mouth, and the red wine had just enough bite to cut through the grease. The fries were salty and crisp, and the salad was crisp and fresh. The chef had included a smattering of palm hearts making it a meal to remember.

It was well after midnight when I paid the bill and dragged myself out onto the sidewalk. The night air had cooled considerably, and a fresh breeze blew in from the mountains. The clinging humidity of the early evening was gone. Thunder rumbled, occasional sheet lightning danced off the buildings throwing a stark light onto the faces of the crowds that were still out in force.

"Mister, Mister," I heard a familiar voice alongside me.

"Mister, Mister you need come meet my sister. She's a birgin, look like a movie star." The ten year old had changed hats. He was no longer hawking merchandise, but was offering his family. I shook my head. "You no like girls Mister? You wan boy?"

"No thanks," I said, hoping he would quit his sales pitch and leave me alone. I felt guilty having just gorged myself while I knew the kid was only looking to make enough so that his family could eat.

"Let me see those watches," I said. "Maybe I will buy one." I was immediately surrounded by a dozen kids, all hawking watches.

"Mister, Mister, look here Mister. I gots better deal Mister. You buy watch my sister go home wit' you. She's a birgin. Very beautiful." A woolly headed kid looked up at me with pleading eyes. My mood changed from heady exhilaration from a night on the town, to a disconsolate pity for the kids of the street. I wanted to run. I wanted to give them something, but I had nothing to give. My mind was whirling, and while my jumbled thoughts tried vainly to think of a way out, the gang dispersed.

"Tomorrow Mister, tomorrow I bring my sister. You see how great she looks and then you decide." An optimistic final sales pitch rang out and hung in the air.

"Tomorrow Mister."

I was determined not to let the incident ruin my evening. My own financial status could hardly be called flush, and there must have been tens of thousands of street kids roaming the back alleys of Bahia. Buying a watch or accepting the offer of their sister would hardly have helped, but for those living on the edge, every bit counts. They were gone for now, swallowed up by the crowds.

I had a night cap on my mind.

I made my way back to the smoky bar and eased through the doorway, clutching my stomach for the amusement of the bartender. The place was packed and the noise level had increased significantly. There was hardly elbow room. The girls had slipped from their perches and were working the crowd, their heavy make-up and caked-on lipstick straining to remain intact. They were already showing signs of a rough evening. I ordered a shot of black coffee and a glass of house cognac. Any residual grease from the meal would not stand a chance against the combination. The coffee and cognac worked against each other with a delightful effect. The coffee threatened to send my pulse rate up through the roof, while the cognac eased into my system, slowly entering my bloodstream, attempting to shut the system down. I felt the tug-of-war in my belly and forgot about the street kids. Someone produced a concertina and the place erupted into song following a scratchy, offbeat tune. I didn't recognize the melody and had no idea what the words were, but the anonymity of a back-alley bar brings out the Pavarotti in all of us. I belted out a few songs reluctantly acknowledging that the cognac was winning the tug-of-war. My eyes smarted and my head thumped.

It was time to get back to the boat.

I made my way back through Pelhourino square which was thronged with people and buzzing loudly. Young mothers were pushing infants despite the late hour. The thunder was closer and lightning flashes more frequent. The harbor was dark except for the anchor lights that hung suspended above the water like glowworms. I couldn't make out which light belonged to our boat, until lightning flash-framed the image for a microsecond. The harbor was bathed in brilliant light, and then plunged back into darkness. We were anchored the farthest offshore.

The elevator was empty·for the ride down, and my thoughts were preoccupied with getting back out to the boat. I needed to find the dinghy, which I hoped was still tied to the dock where I had left it. I had disconnected the fuel line and removed the safety catch, about as much precaution as you can take to discourage theft. The sand was cool and soft, and I slipped my shoes off enjoying the walk along the beach. Part way along I stumbled over two bodies intertwined in the shadows, and heard a low curse emanate from the fleshy folds. My presence did not seem to deter them – a Bahia night has that sort of effect on you.

The dinghy bobbed a gentle greeting, and then, suddenly, a loud clap of thunder rolled across the harbor and the air turned cold and damp. Rain could not be far off. I fumbled for the safety key and connected it to the outboard. The end of the fuel line was lying in the water, so I banged it against the side of the dinghy, hoping to dislodge any moisture, and then dried it carefully on my shirt. My luck with engines has never been good, but this one had yet to let me down. It started on the first pull.

I cast off and headed out.

The inner-harbor was littered with crab pots, and I threaded through them hoping I would not snag one on the propeller. I fought the urge to gun the engine and roar out to the boat. Rain was imminent and I wanted to stay dry. I would be lucky to make it without getting wet, but wasn't worried. The red wine and cognac had left me in a contented frame of mind, and a bit of rain would not have mattered. I motored slowly through the pots until I was well clear of them, and then gently opened the throttle. The engine chugged in response. A brisk offshore wind whipped up an occasional wave which lapped into the dinghy. I could either gun the engine and take a soaking from the waves, or slow down and chance not beating the rain. I decided to gun it; once up on a plane the waves would not be a problem. I opened the throttle – the engine roared, and then quit. The sudden silence was replaced by the sound of wind whipping through the rigging of a nearby boat. I pulled the starter – nothing. I pulled again. Still nothing. I choked the engine and pulled again. Not even a splutter. The breeze had really kicked in and it was pushing the dinghy out to sea. I pulled a few more times putting all my effort into it, still without success. I could smell fuel and guessed it must be flooded. The only thing to do was wait. The night was black, with heavy clouds hanging low. The lights of the city were blurred by a low fog, and their glow barely illuminated the harbor area. We started to drift away from land. Beyond the boats out to sea, it was a black night.

The thunder now rumbled continually, and in the lightning flashes I could make out the lay of the land and the boats at anchor. Our boat was a quarter of a mile abeam of me, and in the darkness I could see her anchor light bouncing in the wind. I found the small dinghy anchor and chucked it overboard. It touched the bottom and then skipped along. We kept drifting. I was starting to feel the first pricks of apprehension. What if the engine did not start? There were no oars in the dinghy. I had left them on the deck of the boat. I jumped up and with renewed vigor, pulled at the starter cord. Still nothing. No sign of life. Not even a spark. I yelled out to the boat, but the wind whipped my words and flung them out to sea like spindrift. Things were not looking good. The mellow afterglow of a good evening on the town was rapidly being replaced by panic. I yelled again, but to no avail. It was no use. I sat cursing the engine and tugged at the starter cord a few times out of frustration, but still no luck. I was in trouble. Our boat was now almost half a mile dead upwind. And downwind? It was a long way away, but Africa was over the horizon. I cursed and kicked and tugged at the starter, and then cursed some more, all without effect. The wind whipped the tops off the waves and flung them into the air. Lightning flashed constantly, followed by loud cracks of thunder directly overhead, and then darkness.

"Fuck," I yelled into the night. "Someone help me please," but my words were lost in the wind. I slumped onto the bottom of the boat and stared at the rubber sides. I was in big trouble and I knew it. By morning I would be miles out to sea. Damn, what an idiot. What a stupid idiot. What a dumb, stupid, stinking thing to have happen. I could taste the frustration in my throat, and feel the sting of salt in my eyes.

The dinghy kept drifting.

It was dark and I was scared. I kicked the engine out of desperation, and then yanked the starter cord one last time. No sign of life. Nothing. I slumped back down, resting my head on my knees and wondered what I would do.

We kept drifting.

I looked out to sea, and thought I saw something, but it was only my eyes playing tricks. Another flash of lightning and I thought I saw it again. Maybe there was something there. What could it be? I waited for the lightening to flash again, but it had stopped.

"Come on man," I moaned. "Give me a break." It was pitch dark, and then suddenly another flash of lightning. Sure enough there was something directly downwind of me. I could barely make out a low platform, and then I remembered what it was. When we sailed into Bahia earlier in the day, I had noticed some fishing boats tied to a platform. It was anchored offshore and the fishermen were using it to gut and clean their catch. I distinctly remembered the smell of rotting fish guts. I hung over the bow of the dinghy, and with both arms paddled as hard as I could. I could not afford to miss it. Between flashes it was pitch dark, and I could only guess where to paddle. Closer now, and then suddenly I smelled it. I was right on target for a perfect landing. I jumped onto the platform and fastened the painter to a metal ring. At least I wasn't still headed for Africa. My heart was beating wildly. I could no longer see the lights of the city, or any of the anchor lights, and our boat, with its crew tucked safely into their bunks, was about a mile away, dead upwind. I sat on the platform and stared into the night, shivering from the cold, and hugging my knees under me.

After the initial euphoria of a safe landing, I took time to assess the situation. It was not good. The platform was covered with fish guts, and the gulls had been by, leaving their calling cards. It was a slippery mess, certainly no place to spend the night. I thought about sleeping in the dinghy, but then things got worse. The skies opened up and rain started to fall. It came down cold and steady, drenching me. I had to stand to avoid the rivulets of fish guts that ran along the cracks. Visibility was down to zero as the rain hammered on the platform and turned the water white. The good news was that the wind had subsided, but it did little to comfort me. I had only one option. I was going to have to swim for it. It was not a great thought, but I had no choice. I was not going to stay on the platform, so I removed my shoes and jeans, and slid into the water. It felt warm after the cold wind and rain. I took one last look in the direction of land, let go of the platform, and started swimming.

I am a reasonable swimmer, but not cut out for long distances. At school I had tried hard to avoid swimming lessons, preferring to be on top of the water, rather than in it. The rain had smoothed the surface, and with the wind down I swam easily, saving my energy. On a calm day in the middle of summer I would have had difficulty covering the distance, but this time was different.

I had to make it.

I felt the adrenaline squirt into my bloodstream. I guessed where the boat was, and swam in that direction. Salt water splashed into my face choking me, but there was nothing I could do except push towards my invisible goal. I settled into an easy rhythm and let my mind wander. I thought about the street kids and wondered where they would shelter from the rain. I thought about the hookers in the smoky bar and wondered how much they made on a good night. I thought about the priest in the old cathedral, the glow of the candles around the alter and of the little girl with her rosary. My mind drifted back to my childhood and I thought of my family back home in South Africa. I wished I had never left. I cursed myself for eating too much dinner, and wondered what effect the cognac would have on my stamina.

The platform and dinghy were no longer in sight. It was black all around. Most of the town would be asleep, and any lights would be blanketed by the rain and fog. I was enveloped by darkness, and felt the first fingers of panic starting to rise again. My gut constricted. I felt my body tighten with tension.

"Cool it," I told myself. "Don't panic." I knew that if I panicked it would be all over. There were no landmarks for bearings – only the direction of the wind. I was certain that the boat was upwind, or at least I assumed it was, so I kept swimming into the waves, fighting to keep panic under control.

I had been swimming for over an hour when the rain suddenly stopped, and the visibility improved. I could see lights ahead, but had no way of knowing if they were anchor lights, or something on land. At least I knew that I had been swimming in the right direction. My legs felt heavy, and my arms burned with each stroke. I strained to see the anchor light again. I needed something to work towards, a beacon of hope. Anything to keep me going. But it only got worse. The wind picked up again, and it blew straight into my face. The waves increased, and an occasional rogue swamped me. I choked and gagged and tried to keep swimming. There was nothing else to do. With only an occasional flash of lightning, I had no means of figuring out which direction I was heading. I swam and spat water, and swam some more until I felt a numbness seeping in. I felt like quitting, and wondered if I would die. I stopped to tread water, and thought about drowning. Would it happen quickly? Would it hurt? Would they ever find my body? I felt overwhelmed and terrified. I was going to drown. My heart raced and my breathing got shallow and uneven. I was going to drown all alone in a black sea on a cold night. I wept tears of frustration and sadness.

And then I thought of my stint in the army and how the corporals had made us grovel. How they pushed us until we bled. How they humiliated us and tried to break our spirit. How they wanted us to quit. I hated the mean spiritedness. I resented the humiliation and vowed I would never be broken. I never gave up. I never quit. You learn that there is such a thing as mind over matter. You learn to never give up. You never let them see you sweat. You never quit. You don't quit, you don't quit, you don't quit, don't quit, don't quit. The words became my mantra and I started swimming again. I pushed on towards the boat. Don't quit, don't quit, don't quit. For a moment I thought I could see the anchor light, but it was only my eyes playing tricks. Don't quit, don't quit. I couldn't stop – stopping would mean drowning. I had to keep pushing. There was that anchor light again, this time I was sure of it. It wasn't my imagination. It had to be the boat. My arms burned and my legs floated behind me. I had no more strength. Waves kept breaking into my face, and I coughed, and spewed, and choked, and cried. It had to be the boat because I couldn't go any further. "Don't quit," I yelled into the night, "don't quit," but I started to sink, and a warm wonderful feeling came over me. The pain seeped from my arms and legs, and was replaced by a floating feeling. A wonderful floating sensation. "Don't quit," I repeated silently, willing my body to keep going. The pain rushed back, and I ached from the effort. I longed for the floating feeling, but my mantra kept me going. Don't quit, don't quit. The wind howled and the waves swamped me.

I kept swimming.

Then out of the howling I heard a different sound. A familiar sound. The sound of halyards clanging against a mast. And then I saw the light again. This time I was sure that it was a light and not my imagination. It was an anchor light. It had to be. The dark outline of a hull loomed in front of me. Don't quit, don't quit, my mantra chanted. Never give in. Never give up. I could just make out features on the deck. Keep going for few a more yards. Don't give up. Don't give up. Please don't give up. Don't quit, don't quit. I longed for the painless floating feeling, but the light was getting stronger.

I was almost there when I remembered that the deck was five feet above slippery and sheer topsides. How was I going to get on board? I tried yelling, but my voice was only a feeble groan.

"Please someone help me," I begged. I swam the last few

yards and noticed a line hanging over the side. Something to hang onto. Something to pull myself on board with. Maybe I was going to make it after all. With a final desperate push I made it to the boat and clung to the rope. It was all I could do to hang there.

"Someone please," I croaked. "Someone please." The wind whipped my words and flung them into the black night. I felt my arms weakening. I tried to pull myself up, but they were too weak and started to shake from the effort. I lunged at the side of the boat kicking with my feet hoping to wake up the crew, but the noise only echoed in the empty lazarette. There was no way I was going to wake anyone, and no way that I could hang onto the rope all night. I felt myself slipping and cried bitter tears.

"Why now?" I cried. "Why after getting this far? Why can't I get on board? Someone please." I felt a welcoming warmth and relaxation seep slowly up my legs, gently caressing my lower body. Maybe if I rested a while I would be okay. The warmth was soothing, and I slipped back into the water. Perhaps it was too much for me. Perhaps I should let go. My arms were being caressed. They felt warm. My neck was being massaged gently. My whole body felt warm and limp. I closed my eyes and wondered why things had to come to an end in Brazil. Why now? Why there? I had no answer. I felt the warmth reach my brain and it felt good. Warm and spongy and gentle and very relaxing. My hand slipped on the rope, and I started to float.

***

It had been a raucous night on board – so much for everyone getting to bed early. They had started with caiparinos, a Brazilian staple with a bittersweet taste and a kick guaranteed to flatten the most seasoned drinkers. A case of Brahma Choppes had been drunk, as had a couple of bottles of wine. Pete had stuck with beer and had finished his fair share, but a six-pack leaves very little room in your bladder. Long after you go to sleep the residual beer continues to filter down, and it was the persistent pressure that woke him at 3 a.m. He lay in bed for a while listening to the wind rattling the rigging, and felt the boat tugging at the anchor rode. Someone had forgotten to lash the halyards, and one of them was banging against the mast. He tried to ignore the noise and the load in his bladder, but it was no use. The monotonous clanging was keeping him awake. He would have to get up and tie off the halyard.

The cabin was dark as he made his way through the saloon and out the companionway. He hung under the dodger for a few moments, waking up slowly and taking in the night. It had turned cold and he shivered in the damp air. First the halyard. He dashed onto the foredeck grabbing a sail tie as he went, and lashed the halyard out away from the mast. It was suddenly much quieter. And now for a quick pee before hopping back into the warm bunk. He wandered aft to the transom to relieve himself. The deck was wet. It had been raining and the night was very dark. The town was asleep, and out to sea there were low thunderclouds scudding across the water. "We probably won't be leaving until the weather clears," he thought. "I wonder where the dinghy is? Must be tied up on the other side of the boat." Pete noticed a line hanging in the water, but didn't bother with it. "That's funny," he thought. "No dinghy." He checked both sides and decided that I must have stayed ashore for the night. Probably too wet to make it out to the boat, or perhaps I had found love in one of the bars and was sleeping ashore. "Oh the bunk would be so warm," he thought. Just to slip back into bed and listen to the wind blowing outside. It was much quieter now without that incessant banging. But what was that other noise?

Something faint and far away like a gull wheeling on the strong night winds. He knew that gulls land during bad weather, and wondered what the sound could be, but the lure of the bunk was too much for him. Pete slid down the companionway and padded back to his cabin. His sleeping bag was still warm and he pulled it up around his neck. "What could that sound have been?" he wondered. There was something disturbing about it. He wondered if he should take another quick look around the deck, but didn't feel like getting up. Pete lay there for a while listening to the night noises. Sleep didn't come. The sound had disturbed him. He would have to get up and take a look outside.

On deck the wind still whipped through the rigging, but above the sound there was another noise. A plaintive cry for help. He dashed aft to the transom, and in the water could make out a shape. A human shape. Pete is a big guy. He reached down into the water and grabbed me by the scruff of my neck dragging me through the lifelines.

"What the hell do you think you are doing?" he yelled. "This is no time for a swim." I looked up into his big, soft, brown eyes, and wondered if all the angels in heaven spoke with a South African accent.

MEETING NANDO

*"Men play at tragedy because they do not believe
in the reality of the tragedy which is actually being
staged in the civilized world."*
Jose Ortega y Gasset

Memories fade with time, and if it had not been for some clear-air turbulence over Brazil I might have forgotten my long swim in Bahia. I had shoved it to the back of my mind and left it there. A decade and a half later I was flying back to South America, and heard the captain tap back on the engines as we hit the first bump. The seatbelt light flicked on and some of the passengers stirred, but it was three in the morning and most of them dozed off again. I glanced at the television monitors. The screen showed our current position, with a thin red line snaking back across the Caribbean Sea and up the eastern seaboard of the US, all the way to New York. The monitor paged to the next screen and zoomed into a close-up of our location.

We were directly over Bahia.

I lifted the window shade enough to see the lights flickering below, and the memory of the swim came back with a rush.

South America does indeed have a distinct throb. It is the pulse of the people and it connects where you expect it would. Right under your diaphragm. The moment we landed in Argentina I felt its familiar tug. The bustle of the airport was distinctly Third World with a haze of cigarette smoke permeating every sense. The seemingly chaotic immigration procedure resolved itself, and I was ushered into the transit lounge awaiting my connection to Uruguay. It was still early, but the bar was crowded with old men sipping strong black coffee and drinking brandy. I believe in the old "when in Rome..." adage, and joined them. *"Uno café solo con Cognac,"* was the extent of my Spanish. It was sufficient to impress the bartender, and he placed a shot of rich black coffee in front of me, poured a tumbler full of brandy and placed the bottle where I could refill at will.

I propped myself on a stool at the far end of the counter and pulled my laptop out of my backpack. The conversation lulled momentarily at the sight of it. The familiar tug-of-war between coffee and brandy took over and the sights and smells of South America surrounded me. It was a good place to write. The bartender was amused and refilled my snifter. I felt like Hemmingway.

An extraordinary thing happens as soon as the first sentence is set down. Your memories, which have long since faded into the recess of your subconscious, are suddenly reawakened as your thoughts hit the page. The words act like keys, and the little boxes where memories are stored start opening. I was well into recounting the atmosphere of Pelourinho Square when my flight to Montevideo was called, so I packed up my laptop, paid the bill and made for the gate. We landed in Uruguay twenty-five minutes later.

I love the bus ride from the airport in Montevideo, to Punta del Este. It reminds me of Africa. Goats and chickens scatter as the bus passes through small towns and winds along the waterfront, until the bright modern buildings of Punta come into view. The big flashy homes and trendy boutiques contrast starkly with the surrounding poverty. Faded billboards hawk American dreams with a Colgate smile. The gap between rich and poor is made more apparent when the two standards coexist side by side. The poor appear poorer, the rich richer, and the middle class nonexistent.

Ten miles before you arrive in Punta, you pass Casa Pueblo, a sprawling mansion built into the hillside overlooking the ocean. It is owned by Uruguay's most famous artist, and he has made the home his life work. The hundred rooms are interconnected by tunnels, bridges, and an occasional walkway, and the result is a magnificent, if somewhat eccentric edifice.

On my first visit to Uruguay in 1979 I was invited to a cocktail party at Casa Pueblo, to meet "some interesting people." Finding the front door was not easy, but once we managed to locate it we were met by our hostess and taken through a maze of rooms and tunnels until suddenly we were on a balcony overlooking the bay. It was one of the most beautiful sights I have ever seen. The whitewashed features of the house accentuated the graceful architectural lines, and contrasted with the blues of the water, and the faded green of the surrounding hillside. There was a light wind blowing off the water cooling the evening air. The other invited guests were a mixture of Punta socialites and transient sailors. It made for some interesting conversation.

Cocktail parties bore me, but this one promised to be different. I noticed a good looking man sitting apart from the crowd. He was wearing riding chaps and a thoughtful expression. I watched him sip his wine and followed his gaze out towards the horizon. There was an old tramp steamer heading up the Rio de la Plata towards Montevideo. I picked up my drink and sauntered over to talk to him. At first he did not seem to want company, and I was about to leave when he asked me to stay.

"Those old steamers remind me of a bygone time," he said. "I think I would like to have lived back then." His English was perfect, but laced with a rich South American drawl. There was a wistful expression on his face, and a faraway look in his eyes.

"I'm sorry," he said, extending his hand. "My name is Nando Parrado." I took it and felt his strong handshake. There was something familiar about his name, but I couldn't place it. "Still I am not complaining," he continued. "I have been very lucky in this life."

"How so," I asked. He looked me in the eye and told me it was a long story, but didn't offer to elaborate, so I moved on to other topics. His name was very familiar, and I wondered if we had met before. He asked me about sailing. He asked what it was like out on the ocean and asked if I was ever scared. I had just turned twenty-one and was invincible in my naiveté. I told him that the ocean did not scare me. He smiled at the comment. His questions were probing and I was enjoying the conversation all the while feeling sure that we had met somewhere before. I knew it was unlikely, but his name was so familiar. Nando Parrado. Nando Parrado I thought to myself, and then I remembered. The Andes air crash. The Uruguayan soccer team whose plane had crashed in the Andes mountains. The survivors ate the dead, and they were eventually rescued months later after one of the group walked out and found help. That man was

Nando Parrado. I wished I hadn't told him that the ocean didn't scare me. Nando knew instinctively that I had recognized him, and he waved his hand at the house in an expansive gesture and said;

"All this wealth, and there are times when even money is not enough." The Uruguayan artist had put up huge sums of money for a search and rescue effort, but they had failed to find even a trace of the plane. His own son had been on the flight.

During my stay in Punta I got to know Nando. He came sailing with us, and took me horse back riding along a stretch of deserted beach. I learned from him the importance of living life to its fullest, and I learned about the strength of the human spirit and its natural instinct for self preservation. He told me about never giving up hope. He told that if I ever found myself in a crisis situation to always look at the immediate picture and not at the big picture.

"The big picture can overwhelm you," he said. "You need small victories to nurture your spirit and to give you the strength the go the distance. It's the exact opposite of life where you need to concentrate on the big picture – the small details can sometimes trip you up."

He also gave me some very useful advice which I hope I never have to use. I was surprised to hear him talk about it. He told me that if I was ever in a similar life and death situation, that I should not hesitate to eat the dead. Those among them that had waited until all their rations were gone before eating human flesh, had died of protein poisoning. It seems that human flesh is very rich in protein, and when supplemented with other foods can be a useful source of nutrition, however when eaten alone on an empty stomach, it can be deadly. It's good advice which I hope neither you, or I, have occasion to use.

A month after leaving Uruguay we stopped in Bahia. Nando's words were with me on my long swim for the boat. "Always look at the immediate picture and not at the big picture," he had said. "The big picture can overwhelm you." I might have not made it if I had considered the big picture. It was just to far for me to swim. It was good advice – life's lessons are found in strange places.

My bus pulled into the terminal in Punte. I would be in the city for two weeks. Each evening after dinner I grabbed my laptop and headed down to a café on the waterfront. I ordered coffee and brandy and wrote these two stories while the alcohol and caffeine fought their familiar tug-of-war in the pit of my stomach.

PETE

*"Please do what you can to take care of the beau-
tiful places. Hear the deep song of the land and
sea. For if we lose this place, this sense of place,
then we shall truly have nothing at all."*

Rachel Carson

The bushmen have a time-honored saying which we would
have done well to heed. "When you see a black rhino," they
say, "you first look for a tree to climb, and then you look at the
rhino." It was good advice, but we were already half way across
the clearing before it sank in. This was not a good idea. Behind
us was open veldt with the nearest tall tree at least half a mile
away. Ahead, the rhino and short scrub. We were on foot with
nowhere to run, and worse still, nowhere to hide.

The pair of black rhino which we spotted earlier had wandered into the thicket ahead and were now out of sight. Pete

had been the first to spot them. They had been browsing along the edge of the scrub blending perfectly with the bush. It was only the sudden movement of the tick birds that caught his eye. A quick flash of white.

"Shh. I think I can see them," Pete said. He kicked at the dry ground with his good foot and watched the dust drift lazily towards us. He kicked again to be sure and then nodded to me. We were still downwind. "We had best be quiet from now on," he said.

***

It had been a long dry winter on the northern plains of South Africa. The sun had sucked the moisture out of every living thing, leaving behind a parched and harsh earth. It had been easy to find game. Just find water and wait, and soon you would see all there was to see as the heat of the day forced all living creatures from their cover out into the open to drink. We had seen just about everything we wanted to see. Everything that was, except for one of Africa's most dangerous animals, the black rhino. They are the ones with short necks, usually found browsing on the inch long thorns of acacia trees. A diet akin to eating nails. They are also the ones with bad attitudes, probably caused by a belly full of undigested thorns. Their gnarly nature and impressive stature combine to make them dangerous and unpredictable. A black rhino would sooner pulverize you and debate the issue later, than bolt off into the woods to avoid a conflict. What they lack in eyesight they make up for with their keen sense of smell and robust personality.

We were psyched to see one before leaving the reserve.

A game guard told us of a pair that could usually be seen browsing near the old airstrip.

"Get up early," he warned. "You're most likely to see them

at first light." So we pitched camp in a stand of acacias and rose before dawn. The air was soft and smelled of dust and dew. It was quiet, the time of day when there is a truce between the hunters and the hunted. The terror of night slowly gives way to

We pitched camp in a stand of acacias
and rose before dawn.

the cruel reality of day, but for a while in the predawn, all is still. A power vacuum exists.

As dawn broke in the east we watched the sun rise over the scorched bush. It rose quickly like a giant orb breathing fire.

The drops of dew that had settled on the grass overnight quickly evaporated into instant humidity. The soft landscape became hard and harsh as the lushness of the early morning disappeared with the first fingers of warmth. Another hot African day had begun, and as we hiked across the clearing I felt the first drops of sweat run the length of my spine.

"Let's stop a moment," Pete whispered. "Let's see if we can see them." We stood motionless for a few minutes and then suddenly, another flash of white. The tick birds were having a feast. "There," Pete said, "just behind that clump of bushes." I squinted into the nearby bush looking for a sign of life, but saw nothing. Pete kicked at the ground again and watched the dust rise and drift lazily on the gentle breeze. We were still down-wind.

"Let's get out of here," I whispered. "Those guys will fucking pulverize us if they see us." I had spent many years in the bush and knew better, but the others were intent upon seeing the rhino and said nothing. Pete tilted his head to the left and nodded in the direction of the tree line.

"There," he said. "There's the other one browsing in the scrub." I followed his gaze and could just make out the bulk of the other rhino. It was less than a hundred yards away, standing motionless. I wondered if they knew we were nearby. I glanced over my shoulder at the open veld with cropped sun scorched grass, and remembered again the bushmens' warning. "First you look for a tree to climb…" In the distance the shrill cry of a fish eagle broke the silence. It was the only sound other than my heart pounding and the gentle clanging of Pete's crutches.

"Pete," I said. "Let's get out of here." He smiled and said nothing. This was his environment and I knew that he would not take any chances, but still I had an uneasy feeling in my gut.

"OK, be very quiet now," he said. "Let's see if we can get a bit closer." Pete rose slowly and started to make his way forward. His crutches clanged quietly as he placed them carefully to avoid cracking dry twigs. I felt sweat leak from every pore and followed a few steps behind. The others fell in beside me. Both rhinos had blended back into the bush and were nowhere to be seen. We crept forward, a small party of rag tag tourists and one man of the bush. The heat was becoming more intense. I watched beads of sweat run the ridge of my nose, coast slowly towards the ground, and splatter on the dry dust at my feet. The moisture was instantly swallowed up by the parched dirt. We moved carefully towards the edge of the clearing; closer to the rhino.

Pete stopped and knelt down. He picked up a handful of sand and let the dirt filter slowly between his fingers. I watched the dust drift towards the rhino. He glanced at me and then picked up another handful. The dust drifted in the same direction.

"Shit," was all he said. The wind had changed direction and we were now upwind. The rhino would pick up our scent in an instant.

"Now what?" I asked.

Pete smiled at me and said, "this should be interesting." I heard the fish eagle call again from its far away perch. It was hunting for breakfast. My shirt clung to my back in a wet soggy mess.

***

As I followed Pete, I thought about man's ability to adapt to his environment. He did not seem to notice the heat like the rest of us. He was hardly breaking a sweat, but then a decade in the African bush conditions you for hot weather. Since graduating from university with a degree in wildlife management, Pete had been living in the wild as part of the conservation crew at Pilansberg Nature Reserve. The elements had tempered him, and with time he gradually became part of the African landscape, equally at home with its harsh climate as most of the four legged creatures that roamed the plains around him. His skin color blended perfectly with the hard earth, and his hair waved like the bleached thatching grass found abundant across the South African low veldt.

We were always close. I think it was the streak of individuality seared into each of our personalities that drew us together. Perhaps it was our mutual love for the outdoors and adventure. Maybe not, though. Maybe it was just that as his younger brother he could use me as a fall guy for his many schemes, and so he tolerated my presence. Whatever it was, we spent a lot of time together – and we got into a lot of trouble. In the mid '70s, while climbing in the mountains, we crossed the line from innocent fun, to dangerous stupidity, and our lives were forever changed.

As children we spent our family holidays in the Drakensberg mountains, hiking, swimming and soaking up fresh air. It was an environment that tested us and allowed us to play outside of our comfort zones. While the mountains are wild and rugged and unspoiled, they are still a relatively safe place for kids to roam. If you hiked up towards the escarpment and lost your way, you simply hiked back down again and chances are you would end up back where you began. Our parents let us roam freely and I would follow Pete on many treks into the bush. Sometimes he tolerated my presence; other times he would stride on ahead purposely trying to lose me. I would always hang on grimly trying to match his stride, always a few steps behind, never giving up.

I remember one particular walk. We hiked higher and higher into the hills while the fog rolled down to meet us, and before long we were engulfed in its wet embrace. Pete has an uncanny sense of direction and strode ahead while I staggered along a few paces behind, trying to keep up. The damp fog mingled with my tears. My legs were bloody from branches and thorn bushes that had scratched them, and my entire body was tired and sore. Occasionally we would startle a secretary bird and it would take flight, rising like a ghostly apparition and disappearing into the mist. We kept going up and up until late in the day when Pete finally turned around, and I gratefully followed him back to camp. It was dark when we arrived back at the hut, and I discovered the first adrenaline high of my life. The exhilaration of a tough day, combined with the relief of making it home safely, made for a heady feeling. Like most addictions, however, getting the high soon became harder, and after a while it required more daring and trying feats to achieve the same effect. For two barefoot African kids, the mountains were a veritable feast where we could test our courage and our blossoming climbing skills. It was all fine fun until the day we tried to go straight up.

It was not a high cliff, or at least it did not appear to be that high from the bottom looking up, but then it always seems higher from the top looking down. The day before, we had climbed a much higher cliff in an effort to get above an eagle's nest. Pete wanted to check for chicks without disturbing the nest, so we climbed an adjacent cliff hoping to look down into the nest. The first thirty feet were an easy scramble up the side of the rock, and then a more difficult climb to a narrow ledge which would take us to a vantage point above the nest. We clambered onto the ledge, and with the enthusiasm of two kids on a mission, quickly made our way out onto the sheer part of the cliff. Initially the ledge was quite wide, but it narrowed rapidly, and suddenly we were standing on the verge of a three hundred foot drop. I looked down and was instantly paralyzed with fear. Pete felt it too. We pressed up against the side of the cliff and stood frozen. My mouth was dry and I could hear my breath rasping in my throat. We were in big trouble. Up was out of the question. Down was not an option, and the way back looked like the north face of Everest. Worst still, Pete had been leading on the way out, and to return, I would have to lead. My gut tightened and I could taste the bile rising from my stomach.

I stood frozen until Pete ordered me to head back. There was no way. I couldn't move. My feet were glued in place. Pete kept urging, realizing that if I didn't move, he too was stuck. The urging soon changed to commanding, and then with gentle coaxing interspersed with loud threats, I slowly started to make my way back towards easier terrain. My shallow quick breathing combined with a shot of adrenaline to make me light-headed, and it soon felt as if I was floating. Floating until the ledge widened and the ground rose up to meet us. We were safe again, and the heady feeling of an overdose of adrenaline and fear took over. We flew back to camp; our feet a few inches off the ground. We had conquered our nerves or so we thought, and the rush was incredible.

So, by the previous days standards, this cliff did not look that high. Eighty feet at the most and no chance of a sudden sheer drop. It was late in the day and the shadows were lengthening, but we figured we would be to the top and down again before it got dark. We lingered at the base trying to decide on the easiest way up, and then without much of a plan, started climbing. My bare feet made easy going of the initial climb. I wrapped my toes around small branches and quickly inched higher. Pete was leading, using his longer reach to drag himself up the cliff. His mind was focused on the top and his concentra-

The Drakensberg mountains in South Africa
—a place of many childhood memories.

tion was complete. He seemed to have forgotten the near miss of the day before. I hadn't, and kept looking down. It soon became obvious that I wasn't going to make it. The panicked and paralyzing feeling of the last climb was foremost in my mind and I knew I was going to quit. My legs were weak and my heart was racing. I turned around and carefully made my way back down. At the bottom I sat with my back against a rock and watched Pete continue his climb towards the top. He stopped just below the summit and perched on a narrow ledge resting and contemplating the last few feet. Standing up and reaching as far as he could, the ledge above was a few inches beyond the tips of his fingers. I saw him feel to one side and then the other, but there was no easier way. He looked down at me and signaled with a thumbs up. He looked much more confident than I felt.

"Come on down Pete, let's go home," I yelled up at him. "It's getting late." Pete just grinned and stretched for the handhold again. He was still a few inches short.

"I'm going to make it to the top or bust," he yelled, and I could see that there was no way of talking him out of it. The challenge was taking over and the adrenaline had kicked in.

"Come on Pete, let's go back, it's getting late," I persisted, but he ignored me and continued to search for a handhold. I thought for a moment of leaving him and heading back to camp in the hope that he would follow, but instead decided that I might be able to help him.

It was to be a fateful decision.

A short distance back along the path, was another trail that I was sure led to the top of the cliff. Perhaps I could get above him and see if there was another handhold for him to reach for.

"I'm going to try and get above you," I shouted, and Pete gave me a thumbs up.

I made my way along the path and soon found the other track. It was nothing more than a faint trail used by the hundreds of rock rabbits that made those cliffs their home. I scrambled up through the underbrush, eased along the contour line, and in a few moments found myself on the flat area above the cliff. I tentatively climbed down to a point where I thought he would be, and peered over the edge.

"Hey Pete, can you hear me?" I yelled. "Can you see me?"

"Don't shout, you'll scare the animals. I'm just below you," a quiet voice a few feet below replied. I couldn't see him. I gingerly lowered myself until I could make out the handhold which Pete had been reaching for, and then peered over the edge into his grinning face.

"If you can get a grip on this ledge you'll be able to pull yourself up." My voice had an edge to it. I was worried. I looked down at my brother and he didn't seem the least bit concerned.

"How much above my fingers is it?" he asked.

"About eight inches."

"What if I just jump up and grab it? Is it big enough to hold onto?" I looked at the small ledge just below my feet. It was amply big enough for him to grip onto.

"Sure," I replied, "it's bigger than the one you're standing on."

"OK," Pete whispered, "I'm going to jump for it," and without hesitating he jumped. His fingers found their target and gripped the bare rock.

"Just pull yourself up Pete," I yelled, my voice tense with excitement. I could feel my throat tighten. Something didn't seem right. "Just pull yourself up," I urged.

"I can't. I *can't*." His words hung suspended in the air for a few seconds before sinking in.

"What do you mean you can't?"

"I can't pull myself up. My arms are not strong enough. I'm plastered against the rock and there is no way to swing my legs under me." Pete was right, he was going to have to pull himself straight up using just the tips of his fingers.

"What about going back down again," I said, but I knew that it would be impossible. He'd jumped from a ledge smaller than the one he was grimly hanging onto. There was no way he would be able to land back where he had been. Beads of perspiration were popping out on his forehead and I felt my limbs slowly turning to jelly. The paralyzing feeling was returning.

"Shit, I'm in trouble," Pete yelled. "Take off your shirt and lower it to me." I looked at him in terror. "Take of your goddam shirt," Pete yelled, and I could hear the fear rising in his voice. I looked at his fingers and the tips were white and bloodless.

"Lower your shirt down to me and you can pull me up," he yelled. His arms were starting to shake from exhaustion. My whole body was frozen with fear. There was nothing I could do. I reached for my shirt and pulled it over my head, and when I looked back down, Pete was gone. I stared at the ledge where he had just been hanging and there was no trace. A second later I heard a dull thud from below the cliff. A thud, and then nothing.

"Pete," I croaked, "Pete are you all right?" There was no sound from below.

"Pete," I yelled, "Are you all right?" No reply. I screamed, and the sound of my voice jolted my body from its paralysis. I scrambled up the bank and along the contour line letting the bushes tear at my clothes and skin. I did not feel a thing – my mind and body were numb. I slid down the trail in flurry of dust and broken branches until I found the main path, and when I got to Pete he was lying listless at the base of the cliff. I ran to his side and for a moment I thought he was okay. He was just laying there not moving. His eyes followed me as I knelt beside him.

"Shit," he said. "We really fucked it up that time didn't we?" I nodded trying to fight back tears. "My back is broken," he stated matter-of-factly. "Go and get Dad."

The rest of that day remains a blur. It was dusk by the time I made it back to camp. My father was sitting in his easy chair sipping his first whiskey of the evening. We summoned the camp warden and some helpers, and in the failing light made our way back along the path to where Pete lay. He had not moved. I thought he was dead. There was a thin line of blood running from his mouth and pooling on the grass, but other than that nothing had changed. Pete blinked his eyes open and looked at my father.

"Dad, my back is broken," he stated matter-of-factly. "Be careful how you move me." I was still sure he would be fine until he was rolled onto the stretcher. Pete's screams still ring in my head, as does the memory of the slow trek back to camp, feeling our way along the path by the dim light of a single flashlight. Pete was loaded into the back of a van and transported down a rocky mountain road to the nearest hospital. Each bump and turn in the road sent waves of pain through his body until he slipped into the merciful abyss of unconsciousness.

A year later with calipers fastened on both legs, and crutches for support, Pete was able to inch his way out of his wheelchair and onto the parallel supporting bars. With his physiotherapist encouraging every move, he slowly crept along the bars to the end, and then painfully inched his was back to the chair. It was a slow, tedious process, but then healing a broken body takes time. Healing the mind takes longer, but that process was already well underway. There is something magical about the human spirit that kicks in when it is needed most. As soon as you accept the inevitable, the practical kicks in. Pete quickly accepted his fate and got on with making the best of a compromised situation.

Another year passed before the calipers were ditched and Pete was able to get by on crutches alone. His back had broken at the same spot where he'd earlier cracked a vertebrae on his spine. It was a weak link, and when his spine snapped, it severed the nerves to both legs. A lot of surgery, and a lot of bed time eventually healed most of the nerves to his right leg, but his left leg was ruined and hung withered and limp.

Being confined to a wheelchair brought its own blessing. Instead of heading straight into the bush out of high school, as Pete had planned to do, his wheelchair restricted him to a less hostile environment, and university seemed a good option. He graduated seven years later with a degree in wildlife-management. As soon as he was released from the constraints of academic life, he headed straight for the hills to put it to good use. Throughout it all he had maintained his sense of humor.

***

We sat crouched less than thirty yards from the rhino. They had still not picked up our scent even though the wind was blowing directly towards them. Instead they ambled along the edge of the scrub stopping occasionally to sniff the air. At times they would lift their heads and stare straight at us, but see nothing. I was not worried about their eyesight, it was their noses that concerned me.

"Let's back up and get the hell out of here," I offered, but the others sat motionless, intently watching the rhino. There was something fascinating about our predicament. "Maybe they will move off and not bother us," I thought. They were slowly heading away and I hoped they would disappear into the bush and be gone. A troop of baboons sauntered out from the scrub and came our way, but as soon as they saw us they bolted off, muttering as they went. The rhino meantime had stopped moving away and were heading back in our direction. Pete grabbed

One of Africa's most dangerous and unpredictable animals — the black rhino.

another handful of dirt and watched the wind blow the dust directly in their direction. We were still upwind. They would soon catch our scent, and when they did it would only be a matter of time before they investigated. Our options were limited; none of them promising. The nearest trees were either towards the rhino, or back across the plain. If we moved slowly, we still had time to beat a retreat; instead we remained glued to the spot. The fish eagle called again from its high perch, and my heart continued its pounding in my chest. I glanced behind at the open veldt and watched a family of warthogs scurrying towards us. They are a clear indication that god has a sense of humor, and I smiled to myself remembering the story Pete had told us the night before. We were sitting around the fire camped under a clear sky – a perfect place for story-telling.

"I know many animals more brave than the warthog," he started, "but none more courageous, and none quite so dumb. He is a peasant of the plains, a drab and dowdy digger of dirt. Despite his funny looks, the warthog is not an animal to be taken lightly." Pete's face sparkled as the coals reflected in his eyes and the flames threw patterns on his weathered features. "The average warthog stands just higher than a domestic pig, but that is the only thing average about them. Their skin color is the shade of mud, and their eyes are small, dark, and have only one expression – suspicion. What they see, they suspect, and what they suspect, they fight, and that about sums them up. So..," he continued, "let me tell you about my encounter with a warthog."

His story took place at Manyaleti Game Reserve, a small private reserve annexed to the larger Kruger National Park. Pete had worked there for a few years before moving to Pilansberg. A game guard had spotted an injured warthog, and they decided to track it with the intention of darting the hog to take a closer look at its injuries. Pete set off with two game guards, and by late afternoon they were firmly on its trail. They followed the spoor through thick bush, and knew that they had the right hog by occasional drops of blood smeared on the grass. The tracks led to a burrow with obvious signs of an occupant in residence. Warthogs often commandeer the burrows of other animals, easing themselves in rear-end first. They lie there facing out observing the passing scene. They use their long snouts to pile a heap of fine dust in the hole and wait until their enemies come within range. The dust serves as a smoke screen as the hog launches itself out of the hole in a frenzy of dust and tusks, ready to do battle.

The injured hog had backed itself deep into the hole, and Pete and the game guards carefully approached the den. They were well aware that there was no animal more dangerous and more inclined to fight, than an injured warthog. There were streaks of dried blood on the dirt around the hole, and it was obvious that the hog had been using the den for some time. Pete peered into the dim interior and could just make out a long sloping snout, two piercing eyes, and the glint of two razor-sharp tusks. He balanced on one crutch, and poked into the hole with the other. The warthog shrunk deeper into the den, secure in his burrow, but clearly resenting the intrusion. Pete had read somewhere that the only way to get a warthog out of its den was to shake a piece of paper in front of the entrance. It appears that there is nothing more insulting to a warthog that this flagrant disregard for its privacy, and the shaking paper is more than it can stand.

He sent one of the guards back to the car to fetch an old newspaper, and while they were waiting scouted around for a place to hide where they could get off a clean shot. They would have to hit the warthog square-on with enough force to embed the dart so that the tranquilizer could take effect. Occasionally, from deep in the burrow the hog would snort its displeasure, but most of the time it just lay still.

The guard arrived with the newspaper, and they stood together in front of the den trying to figure out a plan. Pete poked into the hole one last time, but the hog just snorted in disgust and refused to budge. He stepped back from the entrance and motioned to the guard to bring the newspaper. Suddenly with a loud snort and a billowing of dust, the warthog emerged from its cover, head lowered, tusks extended. It fled its burrow and barreled strait at them. Pete turned, and with a reflex action balanced on both crutches, raising his legs up into the air. The warthog ran between his crutches and directly into the game guards. It slammed into the closest guard, knocking him off his feet, flattened the second, and without losing momentum bolted off into the bush leaving Pete still perched on his crutches, and the two guards battered and bruised on the ground. Pete came through without a scratch, the guards were less lucky, and the hog was found dead a week later. It had backed itself into another hole, and died from loss of blood. When they found the body, vultures had picked its head clean, but the rest of the hog was preserved, jammed tightly into the hole, facing out.

I smiled to myself recalling the story, and watched the family of hogs approach. As soon as they saw us they scuttled off with tails in the air and tusks at the ready. Their sudden movement had frightened the rhino and they stopped browsing and lifted their heads, testing the air. The closest rhino moved its head from side to side sniffing the breeze. It was looking in our direction, and took a few steps towards us.

"Just sit still and don't move," Pete warned. We sat motionless and watched the rhino watching us.

"Damn," I muttered, "this doesn't look good."

"Shhh." Pete kicked the sand and looked at me. "Still upwind," he said. "OK here's the plan. If this guy charges we're stuffed. Our best bet is to get out of here and up a tree." He looked at Alice and nodded towards a thorny acacia on the edge of the scrub. She looked back at him in disbelief.

"You want me to climb that tree?" she asked. "It's full of thorns." Even from where we were sitting, I could see the inch long thorns. They were bleached white and razor sharp.

"Alice, you have two choices at this point. Either you can get up the tree and hope for the best, or you can get trampled by a pissed-off rhino." Alice did not need additional coaxing. She stood up and started for the tree. "Move slowly, and if the rhino charges, make a run for it." She crept forward, one eye on the tree, the other on the rhino. Within a minute she was at the base of the tree and looked back at Pete hoping he had changed his mind. He nodded to her to start climbing. "Erin," he said, "you're next." She crept towards the tree, but stopped when the rhino started to move in her direction.

"Don't worry," Pete said. "He can't see you. He thinks there's a problem and is coming to investigate. Just keep moving." Erin made the tree and started climbing. "Okay," he said, "the rest of us might be in the shit, but let's sit still and see what they do." Bill, Roger and I held our collective breaths and didn't move.

A cloud passed in front of the sun throwing a shadow across the bush. Pete grabbed a handful of soft sand and let it slip through his fingers. I saw the dust drift towards the rhino and wished I was somewhere else.

"When he charges, don't move," Pete said. "He doesn't know we're here. He just thinks there might be trouble and is taking no chances, but he will not see us until he is a few feet away." Great suggestion I thought. Five tons of pissed-off power pummeling right at you, and all I had to do was remain still.

Then I remembered. I could make a run for it; Pete had no choice. His only defense was to stand still and hope for the best. The rhino were less than sixty feet away. I was no longer noticing the heat.

"Bill," Pete said. "You try and get up the tree, but move slowly, and Roger," he added, "you follow him. Just be quiet. This guy knows something is up and will trample you if he sees you." They both looked at me, and Bill muttered something about wishing he was still out to sea. He crept slowly towards the tree, Roger just behind, and had covered half the distance when the rhino fixed its radar on them. It swiveled its ears until it homed in, and ambled over to investigate. It did not look pleased. Pete took a crutch and flung it in the opposite direction, and the rhino stopped. Roger stood up and bolted for the tree, and then it all happened in an instant. The rhino charged, Bill followed Roger and Pete yelled at the rhino. It stopped, turned and looked directly at us. Pete flung his other crutch, and as the rhino ambled over to investigate, I took the opportunity to get out of there. I spun around and fled across the plain. The rhino heard me leave, stopped where it was and fixed Pete in its radar. It took two steps in his direction and then charged again. Pete stood motionless as it approached, gambling against the odds that it was only a fake charge. The rhino kept coming, its massive bulk picking up speed, and then ten feet from him, it abruptly stopped. It raised its head, sniffed the wind, and then pawed at the ground.

I kept running.

Pete stood perfectly still.

In the acacia tree, Roger had trampled over Erin and Alice in his eagerness to escape the rhino, and was hanging out on a high limb, scratched and bloodied. He yelled at the rhino who swiveled and looked in their direction, and then slowly ambled over to investigate. The rhino walked towards the tree, stopped at the base, sniffed the wind, and then without further investigation took off into the bush. The other one followed, and from my vantage half way across the plain, I could hear them crashing into trees, knocking down bushes as they ran into the thick scrub.

It took over an hour for the tree climbers to make their way down. It had taken only a few seconds to shimmy up, thorns be damned. They were covered with scratches, and their blood dripped onto the dry earth. I made it back to where Pete was stranded, and gathered his crutches.

"What is it about you and I?" he said. "Every time we are out together we end up in trouble." I looked at him and laughed.

"Another close call. Next time we might not be so lucky." I heard the fish eagle call from its far away perch and saw a speck of white on a high branch overlooking the river. It was still fishing for breakfast. "Didn't you say there was a pride of lions near the south gate that you thought we could get close to?" Pete picked up a handful of dirt and flung it at me.

"Maybe next time," was all he said.

"Maybe not," I replied.

Pete is still living in the bush. He's camped out in Maun on the edge of the Okavango Delta in northern Botswana. It is a part of Africa that remains unspoiled by man's ever increasing encroachment on wild open spaces. A place where the Okavango River fans out to form the worlds largest inland delta. A rare area where desert and swamp dwellers meet, and where the seasonal ebb and flow of fresh water attracts migrating herds of wildebeest and buffalo. It is a place where Pete is at home. When I last visited him there, he suggested we try and get up close to a herd of elephants. I declined. His spirit and optimism are an ideal combination for living in that harsh and uncompromising environment. He is true to himself and that is about all you can ask from anyone.

I'm just proud to be his brother.

TEA WITH CONDORS

"I travel not to go anywhere – but to go."

Robert Louis Stevenson

We sloshed and slid through sphagnum bog until we reached the incline, and then it was the beech trees that gave us trouble. Their tangled growth snagged our clothing and drew blood on uncovered skin. It would be easier once the sun came out, but that would be a while yet. The first light of dawn had cast cold shadows, and I shivered despite the mild air. In another hour the sun would be up and we would be above the tree line. For now, though, it was a struggle.

I watched Julia hack at the undergrowth and carve a path for the rest of us. She swung her machete in a wide arc, chopping branches and clearing the debris with her free hand. She had been at it for over an hour and I noticed her back was soaked from perspiration.

"Let me take over Julia," I said. "Take a break." She moved aside to let me pass, and I took the machete from her. We were edging up a narrow valley gradually gaining elevation. I swung at the brush and pulled the loose branches aside. It was hard work and before long I was sweating and panting from the effort until finally we could see a break ahead. We were almost at the tree line.

"I'll finish up," Julia said. "Pass me the machete." I handed her the blade and stood back. She hacked at the thick bush and moments later we spilled out onto the level area above the tree line. The view was spectacular even though we had not gained much altitude. Below us, thick Tierra del Fuego scrub stretched as far as we could see. In the spring you can look down on a canopy of firebush and see a blaze of red flowers scattered amongst the dull green leaves of the beech trees. Late in the season they give way to the piercing yellow bloom of the calafate, and if you are lucky enough to be there when the calafate bears fruit, you can feast upon its delicious berries. Legend has it that anyone who eats the fruit of the calafate will return to Tierra del Fuego – this was my second visit. For now though it was late winter and the leaves were dry and brown. Patches of old snow lay in the hollows and the ground was sodden.

We rested for a while enjoying the view, and then continued on towards the top. The summit was a further three thousand feet above us, and at times we could see the peak shrouded in fog. There was snow on the south side, but the sheer cliffs below the peak were black rock. It was going to be a difficult climb.

"We need to keep moving," Julia said. "It's a long way to the top and we need to traverse the ice field before the sun gets on it." She was worried about the snow becoming soft and unstable. The air was much colder now that we were out of the trees, but at least the going was easier. There was a low butte ahead; beyond we would be into the real meat of the climb. From our vantage it looked very steep, almost vertical in places. We kept a steady pace conserving energy, gradually making our way higher. The view got more spectacular with each step, and after a while we could see a thin sliver of blue below. We were high above the Beagle Channel and saw sunlight dance on the water. Beyond, the snow covered mountains in Chile reflected the early morning sunshine. It was a magnificent setting in a remote part of the world.

My kind of place.

Tierra del Fuego crouches at the tip of the South American continent. It is a land of high rugged mountains and flat grassy plains, and stretches from the fifty second parallel, south to the brooding island of Cape Horn. It is flanked on its northern border by the Straits of Magellan, the west by the Pacific Ocean, and the east by the Atlantic. The area was named Tierra del Fuego by Ferdinand Magellan who saw columns of smoke rising from the shoreline. "It looked as if the land was on fire," he was reported to have said, and so it became Tierra del Fuego – Land of Fire. The smoke was in fact from the cooking fires of the Fuegan Indians who lived along the shore. They hunted and gathered for food, cooking over open fires and keeping warm huddled around smoldering embers. They are gone now, wiped out by disease brought to the area by missionaries that followed Magellan. Mark Twain had it right when he wrote: "Soap and education. It's as lethal as guns. It just takes a little longer."

As we trudged higher I thought about the Indians, and their melodic names rolled around my tongue. The tall Ona Indians hunted guanaco with carefully made bows and arrows. They used the skins for clothing and shelter, and feasted on meat. The Ona were the most powerful of the tribes and would occasionally attack the others, but mostly they lived in peace, keeping to themselves.

I wondered about the Yahgans who wore little or no clothing and scrounged a living along the shore. They used spears and harpoons to hunt otter, fish and seals, and carried fire with them in their dugout canoes. I smiled at the thought of wooden canoes nosing their way through kelp beds with a blazing fire to keep the occupants warm. The Yahgan men never learned how to swim, and would make their wives anchor their canoes out in the channel. The wives would then have to swim ashore!

The Huash were the oldest of the Indian tribes, and were pushed to the eastern tip of Tierra del Fuego by the more numerous Ona and Yahgan. They also hunted and gathered for their food, and lived in huts made of sticks and branches. I sympathized with the Alacaluf who roamed the northern plains dressed warmly in guanaco skins. They rigged sails in their canoes and navigated the deep waters of the Straits of Magellan. Their practicality should have helped them survive, but all the Indians are gone now, along with their colorful history.

We stopped for lunch alongside a shallow lake, and collected snow to melt for tea. We had packed salami and cheese, and someone produced a box of red wine. I passed on the wine — I was already struggling to keep up and the wine wouldn't help. So far, other than a cutting wind, the weather had been fine and the conditions perfect for climbing, but I had been pushing a desk job for the past year and was out of shape. We had our sights set on a jagged peak with a daunting name — "Dientes de Navarin — The Teeth of the Devil." The peak is the highest point on Navarin island, and reaches beyond five thousand feet.

It frowns down on the small village of Puerto Williams, the southernmost town in the world. Navarin Island is on the south side of the Beagle, and we hoped that the view from the top would reach as far as Cape Horn. We finished lunch and rested

A view of the Dientes de Navarin — a jagged peak with a daunting name.

for a while. My thoughts drifted back to my days at high school.

Geography lessons would have me flipping to the pages of South America, intrigued by the rugged west coast and remote tip of the continent. The Andes mountains run the length of

Chile, dipping briefly under the Southern Ocean before rising again in Antarctica. The land is fragmented; it appears to have been shattered by some giant prehistoric upheaval. The thousands of islands that make up the Chilean archipelago stretch from Cape Horn in the south, almost all the way to Santiago in the north. I imagined protected bays and anchorages' waiting to be explored, and at times I would picture myself as a smuggler hauling goods ashore at night protected from prying eyes by a myriad of islands. Other times I would see myself as an adventurer, scaling the Andes with a team of yaks; mostly though I would be aboard the clipper ships rounding Cape Horn in a fierce gale, heading for the trading grounds of the Far East. I had read all the books by the great sea captains, and knew the tales of dozens of Cape Horn roundings; graphic stories of men and ships beating into the teeth of strong westerly gales. It sometimes took weeks for them to clear the Horn. I wanted badly to be a part of it all and set a goal of sailing around the world by way of the southern capes. I tasted the first bite of adventure in 1982, rounding Cape Horn in a full gale during the Whitbread race that year.

It was a memorable rounding.

***

We had been trailing the fleet in a slow boat. *Alaska Eagle* had not found her stride in the southern latitudes, and no amount of canvas and coaxing could help her. We had to sail the boat on the edge the whole time just to keep up, and would carry our spinnaker long after the rest of the fleet had doused theirs. Careening out of control down the face of massive Southern Ocean waves became a part of our every day. Sixty miles west of Cape Horn we were suddenly becalmed. After days of thrashing across the uttermost ocean, freezing from blasts of arctic air and hanging on for a fast ride, we were going nowhere. The sails slatted against the rigging as the boat lurched in left over slop. The air was damp and heavy. My watch was called up from below to help coax some speed out of the boat, but it was no use. What little wind there was, was bounced out of the sails with each wave. After a while we gave up, dropped the headsail, over-sheeted the mainsail, and put the kettle on for tea.

The view was spectacular.

The water was a cold steel gray laced with streaks of white foam left over from a front which had passed through the day before. To our south enormous clouds over Antarctica filled the sky. They were every shade of dark, and hung ominously waiting for us to make a mistake. We were in dangerous waters where the wind could whip up at a moment's notice, and an errant wave would easily topple us. It had happened before; many times. I lay in my bunk trying to sleep, but the noise of cracking dacron jolted my nerves with each slat. I must have dozed off because I woke a while later to the sound of bubbles running along the hull. There was less than a quarter inch of aluminum separating me from the cold water. I felt the boat heel and heard footsteps running on deck. My bunk was under the main cof-fee-grinder winch, and I could see the cogs spinning as the guys on deck trimmed the sails. The sound of the bubbles turned to a low whine, and the activity on deck became more frantic.

"We need some help up here." A voice called down the companionway. "The wind is picking up." I scrambled from my bunk and pulled on layers of damp smelly thermals. We had been living in a wet world for weeks and salt sores chafed against the rough fabric. The thermals had been soft and warm when we left New Zealand, but they were now crusted with salt and filth. I grabbed my foul weather gear and headed for the hatch. The wind had picked up and we were surfing the edge of

Antarctica's weather.

"Bring up the storm spinnaker with you," someone yelled. "We are going to need it." I found the sail in its bin and dragged it on deck. The air was cold and biting. It was blowing directly

Surfing in towards Cape Horn
with a full gale brewing.

off the polar ice, and was laden with moisture. The sea had turned black, and spray whipped across the deck, saturating everything. I clipped my harness onto the jack-line and dragged the sail forward. The on-watch had set up for a sail change, and were ready. All they needed was the new spinnaker. The foredeck hand was positioned at the end of the spinnaker pole with a spike in hand, ready to release the sail from its guy. He took the tack of the new sail, clipped it to a changing strop, and waited for a signal from the helmsman. I looked back and swallowed hard. Twelve thousand miles of Southern Ocean and I was still not used to the size of the waves. In a very short time the seas had picked up again, and huge swells rolled in from behind. The helmsman would wait until the boat was on a good long surf, and then signal to the foredeck that it was time to spike the spinnaker. Our forward speed reduced the apparent wind, and it was easier to get the sail down.

We waited with practiced ease, and then I felt the stern rise to a big sea. *Alaska Eagle* surged forward and white water streaked past the hull.

"Crack the spinnaker." I heard the helmsman yell, and the foredeck hand spiked the shackle. The old sail snapped free, the man on the halyard dropped the sail, and the rest of us gathered it under the boom. The nylon was cold and damp. We shoved it down the hatch and hooked the new spinnaker onto the old sheets. It was our secret weapon. We had the fabric custom woven for the race, and the three ounce nylon was doubled up in places. There was a thin wire running down each leech to take the shock of a sudden collapse. By anyone's measure it was bulletproof. It set with a loud bang, and the boat groaned under the load. The bow dropped away and we took off down a wave at twenty-five knots plunging into the trough and sending spray flying across the deck.

There was less than forty miles to Cape Horn.

***

Tierra del Fuego is Darwin country. It was here a century

and a half earlier that Charles Darwin and Robert Fitzroy passed through aboard the H.M.S Beagle. It was the first trip for Darwin; Fitzroy was returning to repatriate three Indians he had seized on a previous trip. The youngest Fitzroy had traded for a shiny button from his tunic, and named him appropriately, Jimmy Button. When they returned to England, Jimmy Button became an immediate celebrity. They clothed him (for the first time in his life), paraded him in front of British aristocracy and generally assimilated him into British society. He was even granted an audience with the King. Jimmy Button adapted well to his new celebrity and cultured lifestyle, and soon learned the language.

Fitzroy's motives, however, were not all charitable. He had plans for Jimmy Button. He and Darwin would use Jimmy Button as a conduit to the rest of the Indians. Jimmy would translate, and it would only be a matter of time before the godless Indians became god-fearing Christians. As they sailed up the Beagle, Jimmy Button set eyes on his homeland for the first time in three years. He was wearing a British naval tunic adorned with bright shiny buttons, looking every bit the part of ambassador. Darwin was fascinated with the Indians and wrote: "It was without exception the most curious and interesting spectacle I ever beheld. I could not have believed how wide was the difference between savage and civilized man: it is greater than between a wild and domesticated animal, inasmuch as in man there is a greater power of improvement." Using Jimmy Button to understand the Indians would be key, but Jimmy had other plans. The moment the H.M.S Beagle made contact with the Yahgans, Jimmy stripped naked and bolted. He disappeared into the thick Tierra del Fuego bush and was never seen again.

Their efforts were not totally in vain. The Indians were as

curious about the rest of the world as the rest of the world was about them, and Darwin set about studying their language. He was surprised to find amongst their vast vocabulary of over thirty thousand words, no words to describe a higher power. "They were the original Godless nation," he stated. In their primitive ways he found no signs of civilization of any kind. It has been said that excess production is a basis upon which a society becomes civilized. There must be at least a little surplus food to support a chief, a priest, an artist or an artisan, but the Indians had none of even the simplest requisites. They had existed without any contact with the outside world, and as such were "unspoiled."

They were perfect for Darwin's purposes.

Wandering the cliffs above Puerto Williams, I felt a deep connection with Darwin and the Indians. There is something immensely primitive about Tierra del Fuego. Something in the massive peaks whose sheer presence seem timeless. A rugged wilderness that has escaped man's overwhelming need for more room on an already overcrowded planet. I felt privileged to be there.

***

The radar was the first to pick up land. A small blip slowly turned to a larger mass of interference, and the island of Cape Horn could clearly be seen on the screen. We could not see anything from on deck. Spray washed all visibility away. The air was chilled and I shivered despite the exertion. We were on the edge of control and the boat was veering from one wave crest to the next. I was riding shotgun, standing alongside the helmsman ready to assist if the load got too much for one person. Every fifteen seconds the boat would plunge down a wave and bury the bow in the wave ahead. "This is crazy," I thought.

We had wanted to round the Horn with a spinnaker up, but this was beyond anything we had done before. The boat was groaning from the strain and shuddered as the bow lifted out of the water and shook the foredeck dry. No one said a word.

And then we went over.

A cross sea lifted the stern and flung it sideways, beam-on to the waves and the boat rounded up. I grabbed the wheel and together with the helmsman we forced the bow down, but the damage was done. The spinnaker collapsed. It fluttered in the lee of the mainsail for a few seconds, and then started to fill. As it swung out of the dead air it refilled and exploded. Bits of tattered nylon fluttered to the water, and what was left hung from the masthead like a sorry pennant. We pulled the pieces down and readied the foredeck for a headsail.

It had started to snow.

I looked up and saw a small black cloud directly above us. Horizontal snow swept the deck dusting the crew and rigging. I turned my back to the wind and pulled my foul weather hood over my hat and snugged it tight. The Southern Ocean swirled around us. My fingers were frozen and numb. I checked to see that my safety-line was secure and held on for a wild ride.

"You should see land about ten miles ahead," the navigator yelled up from below. I can see Cape Horn clearly on the radar." The guy is nuts I thought – we can hardly see the bow. Heavy flakes of snow had reduced the visibility to fifty feet.

"This squall won't last for long. It's clear ahead and behind." The navigator was monitoring his instruments placing his faith squarely in the hands of technology. "We should try and get more sail up," he yelled. We dragged a small jib-top from below and hauled it onto the foredeck. It was hard work made more difficult by the occasional wave that swamped the deck.

"Christ I'm sick of this," someone yelled. "Let's get around the corner and out of the bloody Southern Ocean." Just then the corner appeared ahead of us. The squall passed through, the

Spindrift flies as we round
Cape Horn in a full gale.

snow stopped, and dead ahead less than four miles away was Cape Horn. It crouched low in the water looming ominously dark and brooding. A silver light shone through heavy clouds and brightened the water which turned from a dark black to silver gray.

"Cape Horn," I yelled. "I see land." It was our first sight of something solid in three weeks. Its silhouette was just as I had dreamed it would be. High, rugged and windblown. The waves were getting steeper and closer together as the continental shelf slowed the flow of ocean. They were dangerously steep. We dragged the jib-top forward and hanked it onto the forestay. At times the bow dipped below the surface of the water, and we were swept aft by a rush of cold ocean. The wind was at a full gale, blowing horizontal spray and reducing visibility. At times we lost sight of land, but then it would reappear again, right on the bow. I made my way aft to the halyard winch and we started the slow grind of raising the jib-top. The heavy dacron snapped in the wind, but soon the sail was up and sheeted on hard. *Alaska Eagle* surged forward, also eager to make the corner and head north. It had been a tough three weeks, but first we had dreams to fulfill. Dreams of rounding Cape Horn under full sail. Dreams that had been born in the pages of dusty school books. Dreams that were about to become a reality. I was glad the wind was honking – I would not have wanted it any other way.

***

My thoughts returned to the climb. We were approaching the snow field. Beyond was the vertical rise to the summit. The snow had softened, but it was still easily passable. I was feeling totally out of shape and tired. My legs ached and my lungs were on the verge of popping. I had wanted this to be a difficult climb. There was a need for it to be tough. I had a debt to repay and planned to sweat my way to the top no matter what lay ahead. At first we trudged through shallow slush that soaked our boots. After a while the snow got deeper and it took more effort to make progress. I slipped and slid and sweated behind the rest, wishing that I had visited a gym before coming on the trip. The damp chill seeped into my bones, and despite a deep wish to make the top, I thought about turning back. The snow was icy crust in places, and I battled to walk without falling. We were roped together. I was bringing up the rear. Occasionally I would glance at the scenery, but instead of enjoying the view, I was wishing I was someplace else. Somewhere out to sea where I was in familiar territory.

***

As we drew closer to Cape Horn I could see birds nesting in crevices just below the rugged spires that gave the land its ominous outline. Southern Ocean rollers slammed into the headland, flinging spindrift high into the air before crashing back into the sea. We were in close; too close. I shouted down to the navigator who was peering intently at the radar screen.

"Roger, we're in too close. It doesn't look good up here."

"We're OK," he replied. "We're just under a mile off, but there is plenty of water." The land loomed above us. The boat surged down steep seas while we hung on. The wind shrieked in the rigging. We had reefed the mainsail and I wished we had set a storm jib in place of the jib-top. There was too much sail up. The steep seas were making it hard to control the boat. We cut a white swath across the ocean as *Alaska Eagle* surfed towards the corner. The clouds lifted, and for a brief moment there was blue sky above. In the distance I could see sun reflect off the snow covered mountains in Chile. We were looking at Tierra del Fuego; a land that I had longed to visit since childhood. Cape Horn was hard abeam. The whole of South America was above us. The crew were yelling with excitement.

And then it hit us.

I saw the cross sea a second before it slammed into the

boat. The helmsman saw it too, and abandoned the wheel.

"Watch out," he screamed. "Hang on." He cowered in the cockpit bracing himself against the helmsman's seat while the wheel spun madly. The wave lifted the stern and shoved it sideways. We swung beam-on to the following sea, and the next roller broke right over the boat. I felt myself falling as the boat capsized and cold water hit me in the chest. I dropped until my safety line snugged tight and I felt it jerk me backwards. Cold water rushed by, and for a few seconds it was quiet. I realized that I was under water. We had been knocked down badly and I was on the leeward side. Icy seawater found its way into my clothing, and the shock of it against my body jolted me. Almost immediately the boat came back upright again. We careened along the side of another swell, and then went over a second time. I felt another blast of cold water. The boat came back upright, and for a split second I saw Cape Horn dead ahead. We were lurching towards land in a mad dance. The keel was forcing us back upright; the wind was filling the sails and knocking us down. We went over a third time and I swung at the end of my safety harness tether. I could feel the boat groan under the load and heard a loud snap. "Christ," I thought. "We are going to lose the rig." My mind was clear. If we lost the mast we were in deep trouble. We were in too close to land. The boat came upright and I looked aloft. The mast was still there, but the sails were shredded. The mainsail had split in two and the jib-top hung from the end of the spinnaker pole. It had ripped up the leech. The heavy dacron flogged in the wind.

"Watch out!" I heard the helmsman yell. I looked aft and he was back behind the wheel. "I am going to bear off and gybe." I felt the boat shudder and then slowly come back onto course. Land was less than a half mile to leeward. The craggy sides of

Cape Horn were sharp and ominous. I could hear the boom of water hitting the rocks. I heard the deep roar above the shriek of wind ripping though ripped dacron. *Alaska Eagle* lurched forward and surfed the crest of a short steep swell. Even with torn sails we were doing sixteen knots. The boat was back under control and a quick headcount confirmed that everyone was still on board. We looked like wet rats, soaked and shivering, but safe. For now.

I crawled forward and slipped the jib halyard off its winch. The sail dropped to the deck and dragged in the water. Someone had already grabbed the storm jib. We gybed the mainsail and headed for deeper water. The boat swung around, and behind us Cape Horn loomed large. "Thank you," I muttered under my breath. "Thank you." I thought of all the sailors and ships that lay on the bottom directly below us, and thanked the gods' for sparing us. We did not deserve to go unpunished. We had been in too close with too much sail up. I heard another rip and saw the mainsail part across a second seam. It was going to be a long sail repair. There was less wind in the lee of the land, and the seas had flattened. We dropped the sails and set the storm jib. I thanked the gods again, and vowed to one day return to the area to pay my respects.

***

And so I was.

This climb was a personal journey back to Tierra del Fuego. I knew that we had been spared when we deserved to pay a higher price for our foolishness. Cape Horn had not come by its legendary reputation easily. Many had perished on its rocky shore. We had been lucky to get by with a few blown sails and a dozed shattered egos. As we limped along the coast towards the Straits of Le Maire, night fell, and in the darkness I could feel the presence of sailors who had been there before us: the seamen who had plied those waters bringing wealth to the west. It was a dangerous time. A time when men with dreams of a Cape Horn rounding never lived to see their families again. A time when massive clipper ships were tossed about and snapped in two by the force of the waves. We were lucky not to be counted among them. Cape Horn had taken many lives, but for now we had slipped its grip.

***

As we edged up the steep face below the summit, I felt as if I was paying my dues. The gods would be watching. I had vowed to return and was keeping my word. My body ached from the climb, and a cold wind cut through my jacket. There was less than sixty feet to the summit. We shuffled slowly towards the top and then suddenly there was nowhere left to go but down. We sat atop the "Dientes" and rested. The view was breathtaking. To the north the Andes were bathed in sunlight. They stretched up the west coast of Chile as far as I could see. Below, the Beagle snaked its way between Navarin Island, and the southern reach of Argentina. Far to the west I could see the small town of Ushuaia nestled along the coast, and below us Puerto Williams snuggled in the lee of the mountain. There was a brisk wind blowing in the channel and I could see white caps marking the surface of the water. They caught the sun and reflected like diamonds, and I swear I could see Fitzroy and Darwin on a close reach just fetching the Murray Narrows. To the south, heavy clouds obscured the view, but I knew somewhere down there Cape Horn was shrouded in fog.

It was time for a celebration. I looked at Julia sitting with her back against a rock. She collected some branches on the way up and had gathered them into a small fire. A clump of snow was melting in the billy-can.

"We're good for something us Brits you know," she said, and produced two tea bags. "I never leave home without them." She was a good guide and a great sport. Before long the tea was hot and ready for pouring. I leaned back against the rock face, Earl Grey in hand, and contemplated the climb. I was exhausted and we still had to head back down. My return to Tierra del Fuego had been a long time coming. I had been around the Horn one more time since our capsize on *Alaska Eagle*, aboard *Drum* in the 1985 Whitbread race, and renewed my vow to return to pay my respects. The area held a strong fascination for me. The ocean, the Indians, the wildlife, and Darwin. I wondered where Jimmy Button spent his last days.

My tea had gone cold and I tipped the dregs onto the ground beside me. The clouds still shrouded the area south of us, occasionally lifting to allow a fleeting glimpse of the islands, before quickly closing in again. I threw a few dry twigs onto the fire and shoved another handful of snow into the billy. The branches spluttered and crackled as a thick smoke rose from the embers. Far above me a small speck wheeled amongst the clouds and I wondered if there was a smudge on my sunglasses.

And then I saw them. Condors. Andean condors. The most magnificent of all South American birds of prey cruising the thermals high above us. They soared amongst the clouds and then disappeared from sight. I rubbed my eyes, and when I opened them again the condors were right above us, so close I could almost reach out and pluck a feather from their magnificent plumage. Their primary feathers arched with exquisite perfection, gaining maximum effect from each minimal movement. Their beady inquisitive eyes reflected the majesty of Tierra del Fuego, and I knew in that moment that the gods had acknowledged my effort. To the south the clouds slowly parted, and far away in the murky distance Cape Horn stood like a silent sentinel guarding the gateway between the Pacific and Atlantic oceans. I could almost make out huge rollers thundering into the steep sides, flinging spindrift skyward – and beyond the Horn itself a bunch of baby-faced sailors trying to round the Cape in a full gale; spinnaker flying and dreams of adventure fixed on their faces.

BEAGLE CHANNEL CHRISTMAS

Ma roto hoki ora ka pa the korero
If the inner man is refreshed, the
conversation will be agreeable

Maori Proverb

A cold wind blew from the west, whipping up white caps and spraying spindrift high into the rigging. Occasionally a williwaw would plummet down from the surrounding peaks and flatten the surface of the water, lacing the channel with white foamy streaks. I was cold. My fingers were numb from holding the wheel, and the side of my body that faced the wind ached from a constant pelting. It would be a relief to be out of the wet and below where the stove was warming the interior and hot coffee laced with rum awaited. "Not much longer," I thought. I could see the island up ahead, and knew we would be anchoring within an hour. The leeward side was a secure place to spend the night, and the island offered good shelter and good hunting.

There was an abandoned sheep station near the anchorage, and a lot of rabbits.

Forty minutes later we dropped anchor and let out two hundred feet of chain. Skip and Jim ran bow and stern lines ashore

The Beagle Channel on
a typical blustery day.

and tied them off to trees, while the rest of us stowed the sails and put the kettle on to boil. I ducked below gratefully and warmed myself in front of the stove. The cabin was strewn with gear, hung out to dry, but it was cozy and warm and smelled of damp socks and dinner. There were a few colorful decorations adorning the main salon and a calendar counting down the days until Christmas. Home sweet home – at least it had been for the better part of a month. We were living "off the land" in Tierra del Fuego, hunting and fishing for our food and enjoying the remoteness of the Chilean archipelago, as far away from the traffic and jams of modern living as we could be.

The forepeak onboard served as a temporary larder, and amongst the cans and carafes of wine, a half dozen Upland geese were plucked and hung to cure. They were the spoils of the past weeks' hunting. Along with the geese was a slab of reindeer meat that we had traded with the crew of a French yacht, and the remains of a beaver, which we shot on the first day out. Its pelt lay salted and nailed to a board to cure. The yacht's steel hull was immersed in frigid Beagle Channel water, and without insulation it was cold and damp up front so with the forepeak closed it was a perfect place to hang the catch.

"Who's up for a little rabbit hunting?" Skip asked. I was tempted to stay below and sit by the heater, but everyone else was going ashore, so I roused myself and pulled on my Patagonia gear. The fleece was warm and dry from the heater, and felt good. I grabbed my rifle and a hip flask of red wine, and followed the others into the dinghy. There was less wind in the anchorage, and the yacht bobbed in the leftover slop. We beached the dinghy, tied the painter to a tree, and scrambled up through the low scrub. There were signs of rabbits everywhere, but not one to be seen. In a cove near the anchorage a flock of geese hung near the water's edge, but we were not interested in them. We had eaten our fill of geese and had more than enough in the larder. A couple of steamer ducks churned past, but we left them alone. They were too tough and stringy, and too gamey for eating. We were out looking for rabbit, but it was more the fresh air and exercise we were after.

The island was less than a mile across, and its vegetation grew at a fifty degree angle, blown that way by the persistent wind that buffeted the trees and scrub. The prevailing wind is from the west, so all the vegetation leans towards the east. We walked part way around the perimeter, and after seeing nothing to shoot, made our way to the south side to pay a visit to the sheep station.

At the turn of the century there had been numerous stations scattered throughout Tierra del Fuego, but foot-rot wiped out much of the profit, and the idea was eventually scrapped. For a while many station owners left someone to watch over the place in an effort to lay claim to the land which was locked in an ongoing territorial dispute between Chile and Argentina, but eventually they left and the buildings fell to ruin. As we approached the old shearing shed, we were surprised to see smoke drifting from one of the outer buildings. On the opposite ridge a horse and rider appeared. He galloped towards us brandishing a rifle, and in rapid-fire Spanish explained that he was looking after the station and we should follow him to his hut. He said his name was Miguel, and pointing towards a rough looking corrugated iron shack alongside the shearing shed, motioned for us to follow. We climbed over the low stone wall and made our way towards the building.

His place was crude, but comfortable. The front area was strewn with rough hand-hewn furniture, and there was a tattered curtain hanging in the doorway separating the front area from the back. Beaver pelts stitched together warmed the floor. We made ourselves comfortable while Miguel tidied up and rolled himself a cigarette. After his initial enthusiastic greeting, he had not said much. Instead he beamed mysteriously, clearly pleased to have company. His cabin had a few colorful streamers hanging from the rafters, a simple acknowledgment of the Christmas season. Miguel broke some twigs and threw them in the fireplace. He poked at the coals, coaxing life back into them, and before long, the fire was cranking and warming the cabin. The warmth dispelled the otherwise damp, mildewy smell that permeated the place.

Miguel's face was weathered and worn, and shone like an old boot. The cigarettes had discolored his teeth, and the nicotine added a yellow hue to his skin. He told Skip, who translated for us, that he was watching the station for the owner, and would be on the island for nine months before his relief came in the spring. Once a month a Chilean supply ship would stop by with provisions, otherwise he was all alone and isolated. He would supplement his diet by hunting and fishing, and occasionally gathering mussels from the rocks.

Miguel continued with his mysterious manner; he was clearly planning something. He produced a half dozen goose eggs and told us how he had robbed a nest and stolen the eggs. He then cracked the eggs into a bowl and scrambled them with a fork, beaming at us, but not saying a word. I wondered if we were getting an early dinner. He clearly had something up his sleeve. After ten minutes of vigorous beating, he headed into the back room and reappeared with a bag of sugar. He placed the sugar on the table, spooned ten heaping tablespoons into the goose-egg mixture and stirred the contents. I wondered about the mixture, but Miguel was not saying a thing. He returned to the back room and reappeared a moment later with a small bottle of vanilla essence. It was obvious that the essence was part of a precious supply, and he carefully measured a few drops into the cap, and then dumped the vanilla into the bowl. Without saying a word he scrambled the mixture, and then left the bowl and tended the fire. Skip asked him what he was doing, but Miguel just pointed to the Christmas decorations and didn't add any information. He was clearly enjoying himself and the secrecy only added to his fun.

Too long alone, I thought.

It was starting to get dark outside and the wind in the channel had subsided. From the front room of Miguel's cabin there was a clear view of the water. It was no longer being whipped into white caps. Instead the water was a cold blue-gray, marked with an occasional streak of white foam. Miguel lit a lantern, and the soft light cast a content glow on the faces in the room. He was enjoying his one-man show, and I was looking forward to whatever it was he was making. He disappeared to the back room again, and reappeared with a six-pack of Old Milwaukee beer and a grin stretched across his face. Our surprise at this strange appearance of American beer in a shack in Tierra del Fuego, was only heightened when Miguel ceremoniously cracked the top of one of the cans and poured it into the goose-egg and sugar mixture.

It immediately foamed up, and then quickly subsided.

He added a second beer, and then a third. The bowl was overflowing with froth, so he set it aside and returned with a larger one. Skip held the new bowl while Miguel tipped in the slimy mixture, and then added a fourth beer. He was clearly having fun – we were clearly becoming alarmed. It had occurred to us that this might be something we would be required to drink, and according to etiquette, we would not be able to decline.

By the time the sixth beer was added and the mixture stirred, the slimy froth was overflowing onto the table and dripping on the floor. Miguel grinned at us and then disappeared once again. We could hear him digging around for something. I glanced at the others who were staring at the bowl of froth, not saying a word. A moment later Miguel returned with four cups. He blew into them and banged them on the table to dislodge the dust. He then produced a ladle, and with meticulous effort spooned the goose-egg-beer mixture into each of the mugs. There were four of us, plus Miguel, and I wondered who would not be getting the treat.

I hoped it would be me.

It wasn't. He placed the first mug in front of me and handed the others theirs. I peered into the cup. The froth from the beer had subsided, and the drink looked just as you might imagine it would – like slimy yellow egg yolk floating in cheap American beer. Miguel stepped back and admired his work. He pointed to the decorations hanging from the rafters, smiled at us and said, "Chreestmas."

We were not smiling.

"Chreestmas," he said again, and gestured for us to drink up. The sugar had dissolved and the vanilla essence added a sweet smell to the mixture. I took a sip and fought the urge to gag, but the mixture slid down my throat and I was surprised to find that it tasted quite good. I recognized the flavor, sort of. The concoction tasted like eggnog; with a twist. The others drained their mugs and started to laugh. Miguel poured another round. This time we all drank without hesitating, but the mixture was rich and one mug was enough – the second cup tasted more like beer and old eggs. Miguel beamed and then headed into the back room once again. He returned with a cold Chilean beer, popped the cap off and sat by the fire. He leaned back and drank the beer. He looked at us and laughed, drank more beer, and then told us what had happened.

Some French sailors had palmed a case of cheap American beer off on him. He had hoped that they would return so he could play a trick on them, but they never came back. When he saw us sailing up the channel, he decided we were to be his victims instead, and we were. Miguel laughed, then brought out a case of cold Chilean beer.

"Much better," he said.

"Much better," we agreed.

It was late in the evening when we left the hut and made our way back to the dinghy. We had finished the beer and enjoyed Miguel's cooking – stewed beaver and rice. The night air was clear and cold and the Southern Cross was visible directly overhead. We rowed out to the boat and climbed aboard. I looked towards the now calm Beagle Channel where a full moon was reflected on the still water. Another day in Tierra del Fuego was coming to an end. I slid into my sleeping bag and thought of Miguel alone on his hut. It would be weeks before he saw another face, and four long months before he saw his family. He needed a strong sense of humor to see him through his long stay on the island – and a good supply of Old Milwaukee.

SPINDRIFT

"The cure for everything is salt water –
sweat, tears and the sea."

Isak Dineson

I dream of the future in color and the past in black and white, and wonder what it means. Are my memories just fading with time? Have they been purged through a magical filter that weeds out sad and bad times, and replaces them with good ones? Is this a selective process which optimists have refined? Optimists and sailors – a way to forget the cold, wet days and remember only the warm, sunny ones. Perhaps, or perhaps it's just a desire many of us have to reconnect with simpler times; a subconscious need to realign ourselves with the natural rhythm of the earth. A need to sever the shackles of society and sail beyond the reach of land. A need to shed the clutter of life and cast our lot with the wind and waves. When I feel the tug I answer the call and go sailing. I toss my lines ashore and point my bow towards a distant landfall. Before long the traffic and jams of life disappear over the horizon, and I am alone at sea. Alone with the elements. Alone with my thoughts. Alone with the gods.

A gentle zephyr throws shadows on the water. I see the darker patch moving towards me and feel pressure on my sails. My boat heels in response, slowly at first and then with more urgency. Water rushes by. Bubbles murmur on the hull as the boat cuts the surface. Waves slap against the bow. The slap slap comforts me. I hear the creak of the halyards and the moan of the autopilot, and feel at home. They are familiar sounds. They are the sounds of life on board. I scan the horizon for sea life and ships, but see nothing. It is an uncluttered world out here, and a purely honest one, too. What you see is what you get. There is time to think and time to reflect. Time to reconnect. The organized pattern of shipboard life, a clear focus and a defined goal allows my consciousness to flourish. I know that what I do with my time will define not only the passage I am making, but my spirit and character as well. It is an opportunity to look inside my head and see what's in there, warts and all.

My childhood dreams were filled with a longing to escape from the constraints of growing up, and I yearned for the weekends when I could get away to the familiar surroundings of the yacht club. Boats gave me wings, and I craved the freedom and independence I felt when sailing. The moment I cast off, I felt the intoxicating feeling of being in control. On the water, I was in charge of my own world, and the magical properties of wind and waves combined into an addictive potion. These same feelings remain with me to this day, but they are now spiced with the need to be different and a desire to challenge myself; an unarticulated wish to be someone. Crossing oceans and sailing around the world provides the perfect means to realize these goals. It is a place where my neighbors soar and frolic and the highway is endless, and I always return to life on land a better person.

The view from my nav station aboard
Great Circle is outrageous.

Perhaps trying to define the lure of the sea is too difficult, and maybe I am using lofty thoughts to mask more basic reasons. Sailing is fun, and the open ocean is a beautiful place to be, especially in the early morning when the light is soft and the sleepiness from a long night fades into the anticipation of another day on the water. There will be challenges ahead, maybe another day like the one just finished. There will be the vagaries of a new weather system to deal with, and hazards yet unmet. The snow might fly again, and the Southern Ocean may kick up into a gray angry cauldron, tossing spindrift high into the air and flinging it across my bow.

More likely though, low swells from a distant storm will nudge my boat towards the next landfall, the monotonous slapping on the hull gradually connecting with my inner beat. The slap slap washes away the clutter. It washes away the grime and slime from living on land. I begin to notice an eerie silence in my head as the constant rhythm replaces advertising jingles, and my thoughts become clear and unfettered. It's an addictive feeling, far from any of the physical aspects of sailing, yet somehow a part of them. It's the reason I love to sail. As Havelock Ellis once said. "The most beautiful of the arts. It is life itself." For me life at sea goes one better.

A SLEIGHT OF HAND

We have to change our pattern of reacting to experience. For our problems do not lie in what we experience, but in the attitude we have towards it.

Akong Rimpoche

You can see the "smoke" from Victoria Falls from a long way off. When you fly over the area, you look down on dry African scrub until an oasis of green and plumes of "smoke" appear. When you see the vegetation change you know you are above one of the world's most spectacular waterfalls. The "smoke" is a fine mist that rises up from the foot of the falls, and saturates the surrounding savanna. The scrub turns green and lush from a constant watering, and contrasts starkly with the dry bush.

Our pilot banked over the waterfall allowing a clear view of the spray and the thin ribbon of blue that snaked its way towards the edge. The Zambezi river winds through barren veldt, and then plummets three hundred and fifty feet into a deep crevasse. The earth has split apart and swallows the river, sending spray high into the air. The falls thunder to an ancient rhythm and we feel its roar resonate through the thin panels of the plane.

Victoria Falls – the Zambezi winds through thick scrub and then plummets over the edge.

It is a spectacular sight, not easily forgotten. The pilot circled one more time, and then headed for a small landing strip on the edge of town. I was back in Zimbabwe fulfilling a promise that I had made to my seven year old daughter years before. She was named Victoria, although everyone calls her Tory, and when she was small I told her stories of a big waterfall deep in the African bush. It was a place where magic could happen; I promised to take her there.

Our six-seater plane touched down on the runway and taxied to the terminal. I could see the heat rise in waves off the tarmac, and felt the humidity buffet us as soon as the door was flung open. I grabbed Tory by the hand and we climbed down onto the apron. The air was warm and smelled of jet fuel and cooking fires. It smelled of Africa – the old and the new.

I was glad to be home.

We shared a taxi for the trip downtown, and headed towards the village of Victoria Falls, or Vic Falls as its known to the locals. The small town has prospered from the tourist trade, and from the downtown area you can see the spray and hear the muffled thunder of the falls. The striking beauty of the waterfall, cheap accommodation, and abundant wildlife attracts visitors from around the world. They come to admire the crack in the earth where millions of gallons of water plummet each day. Occasionally you can see a helicopter on a scenic excursion, flying close to the face of the waterfall, allowing high-paying passengers a closer view. Mostly though, it is a town of young transient tourists, arriving in beat up pickup trucks and combievans, wearing braids in their hair and sandals on their feet. They come for the bungee jumping and white water rafting, and they come because it is cheap, exciting and spectacular. We would fit in just fine.

Tory and I had been in Africa for three weeks. We had hiked the Drakensburg mountains in South Africa, camped out in Botswana, and felt the beat of the land slowly creep under our skin. For me it was a time to reconnect; for Tory it was a time to discover a world beyond television. A few nights before we had camped alongside a shallow lake, and watched while elephants bathed and drank in the moonlight a few feet from our fire. In the night we heard lions roaring, and the next day saw fresh tracks where a pride had wandered through our camp. Vic Falls would be a bit more civilized. We had a hotel for the night and planned dinner on the town. First though, a trip to the falls was in order. We changed our clothes and headed out onto the streets.

With the exception of the main road that runs through downtown, the streets are dirt and a fine coating of dust had settled on everything. Goats and chickens scratched an existence off the land, and we passed a donkey tied to a post. The dirt was worn in a circle around the post, and the donkey stared at us with indifference. Some of the locals nodded as we passed, and then I heard a voice calling out.

"Hey Mister, you need to change some money? Twelve for US, sixteen for UK." A young boy slid in alongside of us and said again, "I give you twelve dollars for US, and sixteen for UK."

"No thanks," I replied. "I don't need to change money." I felt for the familiar lump of my wallet and kept on walking.

"My name is Jacob, like in the Bible," he said. "When you need to change money you come and see me, OK?" I nodded.

"Sure man," I said. "Maybe tomorrow." The boy winked at us, and slid back into the shadows. I did a quick mental calculation. I had heard that you could get a better rate changing money on the street in Zimbabwe, but twelve for US was better than I had expected. The exchange rate at the airport had been eight Zim dollars for each one of my US dollars. It translated into cokes for a quarter, and cold Tusker beers for seventy-five cents apiece. I still needed to buy our air tickets out of Zimbabwe, and would have to change a pile of cash before we could leave.

"Hey Mister." A different voice called out. "You want to see some carvings?" Another kid slid into step alongside us and lifted his filthy jacket to reveal a magnificent stone carving of an elephant. I looked for a moment resisting the temptation to touch it.

"Oh no, thank you," I replied. "I am not interested." I had seen similar looking carvings in the curio shop at the airport, and thought that I might charge one on my credit card.

"One hundred and fifty Zim dollars only," he replied, "but of course, price is negotiable." I did not want to start up a bargaining session with him, and recalled that the elephant in the curio shop was around two hundred dollars.

"No thanks," I said. He dropped back and offered a final pitch.

"One hundred Mister, and if you are interested my name is Michael."

I wasn't interested.

We walked towards the noise and spray and soon found a sign that read "*Welcome to Victoria Falls.*" Beyond was a path. The roar of water increased as we slipped our way down the wet trail, and it was only when we were right opposite the waterfall that the full effect hit us. It was obvious why the locals had named the place *Mosi-Oa-Tunya*, "the smoke that thunders." The water plunged towards a deep pool at the base of the falls, and thundered as it hit the surface. A heavy spray rose up and saturated the surrounding area. Our clothes quickly became wet, and the trees and bushes dripped constantly. We stood alone at the edge of the gorge, and I let the spray wash over me. It washed away weeks of dirt and years of grime and slime from living in a fast paced world. I could smell the earth and feel the water's thunder resonate deep in my gut. The ancient rhythm was anchoring my spirit. Africa was calling me home. Soon I would be on a plane heading for New York, but for now I was enjoying the moment. I closed my eyes and purged the thought. The bustle

Victoria at Victoria Falls.

of life in America held no appeal.

Tory shivered from the cold. She grabbed my arm.

"Can we go now Daddy?" She was cold and wet and ready to leave. I took her hand in mine and we walked back along the footpath towards the gate. In a few moments we were out of the spray and into the hot African sun. The heat quickly evaporated the moisture from our clothes. At the gate I took a photograph of her standing under the sign.

"Victoria at Victoria Falls," she said. "Can I have another coke when we get back to the hotel?" The ten cent cokes were becoming a novelty.

"Sure Sweetheart," I said. "I'm dying of thirst also." I was looking forward to a cold beer.

We stopped in town to buy some sandals, and as soon as we left the store Jacob slid in beside us.

"Hey Mister, if you need to change money I give you good rate," he said. "What you have? US or UK?"

"US," I answered. "But I don't need to change any money."

"I give you twelve for US and there is no commission." I stopped and looked at Jacob. He looked to be in his late teens, perhaps twenty. The earnest look on his face made him appear younger than his years. His clothes were tattered and he was barefoot.

"Let me think about it Jacob," I said. "Maybe tomorrow."

"Yea Mister," he said. "Tomorrow I see you." He slipped back into the crowd and was gone.

Back at the hotel I sat by the pool writing my diary, while Tory charmed the barman with stories of America.

"It's true," I heard her say, "we have over fifty television channels. There's Nickelodeon and....and lots of others." The barman looked at her in amusement and total disbelief, and plied her with more Coke. I ordered another Tusker and my mind drifted back to the street hawkers. "Twelve Zim dollars for one US dollar," I thought. That would be fifty percent better than the bank. I needed to change a fair amount of money before we left Zimbabwe, and my mental calculator soon figured a savings of over fifty dollars. I pondered the amount for a moment, and then went back to my diary.

It was winter and the days were short. By late afternoon the long shadows of the mopani trees shaded the pool, and the air turned cool. Tory extracted herself from the water, and waved to the barman who was still smiling in disbelief at the thought of fifty television channels. A fiery red sky washed the western horizon as the sun dipped below the tree line. We changed for dinner, and went back into town to find a place to eat.

The village had come to life after the heat of the day, and exhibited all the sights and smells of a small third world town. Dust hung suspended in the flickering street lights, and the street hawkers were bothering the tourists with their persistent sales pitches. They all had some great bargain, negotiable of course. In the distance I could hear the falls thundering. There were other sounds of Africa as the wildlife slipped from the slumber of daylight into the harsh reality of night.

We ate steaks in a street café, and I drank a bottle of South African wine while Tory continued to take advantage of the ten cent cokes. It was our last night in Africa. The next day we would fly to Johannesburg in time to make a connection to New York, and then on home to Boston. I could hardly imagine being back in the US, to crowded sterile streets, and the cut and thrust of day to day commerce. African dust had seeped into my system and slowed my pace to a crawl. It was going to take a lot to jump start me again. I paid the bill and we headed back to the hotel.

"Pssst. Hey Mister, it's me, Jacob." A familiar shadow fell in beside us and continued the sales pitch. "I have the best rates in Vic Falls and no service charge. I see you tomorrow. My name is Jacob like in the Bible." The added inflection on the word "Bible" added surety to his sales pitch.

He was surely someone I could trust.

"Maybe," I replied, then added, "but don't count on it."

We fell asleep to the chorus of wild creatures staking their territory and hunting in the dark. Sometime before dawn I heard the far off roar of a lion. It came from across the Zambezi river, somewhere in Zambia, and I savored the sound knowing that it would soon be traded for the more domestic sounds of my American life. I heard the mad giggle of hyena, and the persistent call of the African nightjar, but the sounds faded as light brightened the eastern sky. We were out of bed early and ate breakfast by the pool. Our flight was mid-afternoon and I needed to change money and buy the tickets. I thought about trusting the street hawkers, then thought better of it and opted instead for a bank. It would be a safer bet. On the way into town a silent figure fell into step beside us. A tattered boy no older than thirteen with solemn eyes and a snotty nose tugged at Tory's sleeve and lifted his jacket to reveal another carved elephant. It was beautifully crafted and I couldn't resist touching it. The elephant was cold and smooth and very heavy, and the boy told me that it was carved from a local stone. The tusks were carved from real ivory and he had removed them and stuffed them into his pocket so that they didn't break. He carefully unwrapped them and they fit neatly into two small sockets either side of the trunk.

"You want to buy elephant?" he asked. "One hundred and fifty Zim dollars only. Very good deal." I hefted the elephant, and liked the feel of it. It was a better carving than the one I had seen the day before, and better than the one at the airport.

"No thanks." I said. "I am not interested," and returned it to the boy.

"One hundred only," he said, and he knew I was weakening. I could almost picture the elephant on my mantle at home, and felt a twinge of pity for the boy.

"Seventy-five." I countered, and he knew that he had me.

"Eighty Mister, and we got a deal." The boy took my money and left. Moments later we were surrounded by half a dozen kids all producing a carved animal, all willing to negotiate, all knowing that I was a buying customer. I felt smug at the deal that I had struck and knew I had paid around ten US dollars for a carving that would reach close to a hundred back home. I had struck a good deal. A skinny kid touting a hippo caught my attention, and after some negotiation I owned the hippo for a little under sixty Zim dollars. I packed both carvings neatly in my backpack.

We sat for a while watching the passing trade. I bought Tory an ice cream which dripped in the sultry heat and before long Jacob appeared and sat down beside us.

"What's your name little girl?" he asked.

"Victoria," she replied. "Like the queen."

"No Mon," Jacob responded. "Like the falls and you are much more beautiful." I knew that he was softening us up, but his manner was gentle and Tory was enjoying the attention.

"Can I get you some sweets?" he asked. Tory looked at me and I shook my head slightly. I saw a momentary flash of sulkiness in her eyes, but she knew better than to accept candy from a stranger. Jacob told me that life in Zimbabwe was difficult since the revolution that changed it from the British colony of Rhodesia, to an independent country. He told me that he worked for one of the store keepers who needed foreign currency to buy goods for his store, and that was the reason why he was able to make such a good exchange rate. His story seemed reasonable, and feeling flush from my recent bargaining sessions, I asked him for his best rate.

"Twelve for US," he said. "It's the best rate in Vic Falls and no commission." I thought for a moment and then countered.

"I need at least fifteen."

"No Mon. I can't do fifteen. If you change over a hundred US, I can do thirteen." Now we were getting somewhere. The greed was starting to stir. I had to change close to two hundred dollars and my savings would be significant.

"I'll change two hundred US if you'll give me fourteen." I was willing to hold out for the best rate. There were plenty of other traders on the street.

"No Mon," he laughed. "I can't do fourteen. It's too much." I decided I would stir him up a bit by walking away from the deal.

"Sorry Jacob," I said, "I have to go." I took Tory by the hand and walked away. I felt smug. I knew that I had him. He would go for fourteen and I would save a bundle. I was glad that I had not changed money at the bank. We wandered in the direction of the banks, and before long Jacob fell in beside us.

"Hey Mon," he said. "I can do fourteen if I give you half of my commission." I had him now.

"OK then, we have a deal." I was anxious to make the transaction before he changed his mind. Jacob smiled at me and I felt good about being able to help him make a little money. The banks always make a ton of profit; this was more direct. This would really be helping the locals, and he seemed to be a nice kid.

Jacob pulled me aside, and in a low, secretive voice told me to meet him back at the place where we had been having ice cream.

"Just keep on walking past there," he said. "I have to stop at the store to get the money." I worried that the store keeper might try and talk him out of the deal, but knew that he had to go there to get the money.

"OK Jacob, but don't be long. I have a plane to catch."

"Don't change with anyone else," he said. "I will see you in five minutes. My name is Jacob, like in the Bible." With that he was gone. Tory looked at me with a quizzical stare, and asked if she could have another ice cream. We stopped and bought one each, and then wandered back towards the hotel. The taxi to take us to the airport would be there to pick us up in less than an hour, and we would soon be on a plane heading home. It was very humid and I could feel small rivulets of sweat running down my back. I was nervous about changing the money, but fourteen Zim dollars for one US was a great deal. I could feel the solid weight of the elephant and the hippo in my back pack, and it gave me confidence. Jacob fell into step beside me.

"I've got the money," he said. "Two thousand, eight hundred Zim dollars." I had already slipped two hundred dollar bills into my front pocket. "Just keep on walking," he said. I thought he looked a bit nervous and might back out of the deal.

"Are you sure you want to change money at that rate?" I asked. I felt a twinge of guilt at the lopsided exchange rate.

"Sure Mon," he said. "No problem. Just up ahead there is a small alleyway. I don't want the other people to see us making a deal because they always hassle me. We can change there." We kept on walking, and at the alley I stopped. For a fleeting moment I wondered if I was doing the right thing. Perhaps I should change at the bank, but their rate of eight Zim dollars was not much. Jacob pulled a wad of notes out of his jacket. They were bound by a few elastic bands.

"Quick," he said. "We change money." I pulled the bills out of my pocket and handed them to him. He counted the money, and then handed me the wad of Zimbabwe currency.

"Wait a moment," I said. "I need to count this." It appeared to be in small bills.

"Count it at your hotel," he said. "Don't worry you can trust me. You know where to find me if it's not all there." I had seen him on the streets since we arrived in Vic Falls, and I knew that I could find him again if we were a few dollars short.

"Okay Jacob," I said, "but I will come looking for you if it's not all there." I suspected that he might have skimmed a few dollars for himself, but was not concerned. Fourteen Zim dollars was a good exchange rate.

"Thanks Mon," he said. "And you Victoria, you are the most beautiful girl in the world."

With his final compliment he was gone. I could feel my wet shirt sticking to my body. We needed to hurry back to the hotel. There was just enough time to pay the bill and get to the airport. I slid the wad of cash into my pocket and wandered out into the sunshine. Jacob was gone. Tory ran ahead and I ambled behind feeling the heat and dust and wishing there was time for one last Tusker. The cash was burning a hole in my pocket. I walked on, and a small knot started to form in my stomach. I knew that I should have counted the money.

"Tory, wait a moment," I said. I stopped under a big acacia tree and looked around to see if anyone was watching. The knot in my stomach was getting tighter. There was something about the deal that did not feel right. Jacob had left too quickly. I pulled the cash out of my pocket and looked at it. The note on the outside was a clean blue 100 dollar bill. I flipped the rest and they were all blue. For a moment I felt better, but when I pulled the bands off, the bills fell open and the knot got tighter. The top bill was also a blue note, but it was for one Zim dollar. I flipped to the next one, and then the one after it. Jacob would be smiling to himself by now. He had just made one hundred and ninety American dollars profit for a few hours work. I had been caught by one of the oldest scams in the book. I felt rivers of sweat running down my spine.

We were going to miss our plane.

***

There is a lot more to this story. My immediate thought was to never tell a soul about the incident. Why admit to stupidity when I didn't have to? After some time I thought about what had happened and realized that Jacob was a master at his game. If he had been born into a different society, he would likely have been another Donald Trump. He was very good at what he did, and I had to admire his skill.

I still, however, had the problem of buying the plane tickets, and Jacob had stolen the last of my money. I already had the flights from Johannesburg to New York paid for, but getting back to South Africa was going to be a challenge. It was eleven-thirty in the morning. Our flight was due to leave Vic Falls at one-fifteen to arrive in Johannesburg in time to make our connection back to the United States. If we missed the plane we would have to wait three days until the next flight, by which time our nonrefundable return tickets would not be worth the paper they were printed on.

I grabbed Tory by the hand and ran back into town. The banks closed at noon, and I had to try and get there in time to get a credit card advance. I had already checked and knew that the airline would not take a credit-card.

Tory was surprised by the sudden rush, but kept up with

me as we ran through the center of town. We arrived at the bank just as a stern looking doorman was shutting the front door. I managed to get one foot inside and pushed it open. The teller looked bored and disinterested when I explained my plight. I would need my passport, she said, hoping that her revelation would suddenly discourage me from pursuing the matter further. It was a few minutes before lunch and she was mentally out of there. Luckily I had both passports on me. She took mine and my MasterCard, and disappeared behind a closed door.

I waited patiently looking at the clock on the wall wondering if we would make the plane. It was scheduled to leave in a little over an hour. She reappeared with the bank manager. I was told that they had to call the US to get authorization on the card, and I would have to fill out some forms. I mentally calculated the time difference, and knew that the US banks would not open for another few hours. I prayed that there was some direct link to a computer that would give us the OK. The manager escorted Tory and I to a small airless room and produced a long questionnaire to fill in. It demanded all sorts of irrelevant information. I scribbled as quickly as I could, and handed the forms back. It was almost twelve-thirty. We sat waiting, the doorman shuffled his feet noisily, the teller looked bored and hungry, and the clock ticked loudly, taunting us.

An eternity later the manager reappeared, beaming. He had my money. He shook my hand and told me that it was the first credit card advance his bank had done. We took the cash and bolted outside in time to grab a cab. First a stop at the hotel to get our bags, and then full speed to the airport. I now speak from experience when I tell you that there is nothing more frightening than a Zimbabwean cab driver with an incentive for haste. His car was more than twice as old as my daughter, and badly in need of repair. Bits of it were breaking off as we sped towards the airport dodging donkeys and goats, and finally with its last gasp, we pulled into the passenger departures.

The place was deserted.

A couple of flies buzzed noisily, but that was all. The airport is small, and out of the window I could see a South African Airways jet taxiing on the runway. I banged on a door that said "Customs Only" startling the on-duty customs man who was about to nod off to sleep. I explained that I needed to catch the plane. He jumped to his feet, yelled into his walkie-talkie, and before I knew it, out the same window, I saw the plane stop, turn around and return to the gate to pick us up. We paid for our tickets, cleared customs and boarded. It was the first time a jet had come back for me. As we took off I looked down and saw the "smoke" rising from Victoria falls, and the thin blue line snaking its way through dry African veldt.

A LONG DAY

" Do not go where the path may lead; go instead where there is no path and leave a trail."

Ralph Waldo Emerson

Our truck was packed and ready to go. I had changed the oil the night before and stowed most of our gear in the back, covering it loosely with a tarp. Erin was inside fixing a thermos of coffee and making sandwiches for breakfast. Her footsteps were soft on the tiled floor, and the pale early dawn light cast gentle shadows across her face, softening her features and giving her skin a warm glow. She looked beautiful with her thick auburn hair pulled back and fastened with a childish hair clip. She was not a morning person, and I knew that I would be better off outside until she was good and ready to go. I started the truck to let it warm up and checked one more time to see that our gear was secure. The ride to the "wild coast" was long, most of it on bumpy dirt roads, and our gear was sure to be jolted loose if not properly secured. The tarp would help keep the dust out.

I wandered around the side of the house to the back, and sat on the edge of the lawn looking over my father's small avocado orchard. The ground sloped away from the house, and the orchard ended up against the woods that abutted our land. In the trees vervet monkeys swung with wild abandon, poaching onto our property and swiping an occasional avocado before scurrying back to the relative safety of the trees. If they weren't such pests it would have been amusing to watch their antics. Instead I picked up a stone and flung it in their direction. The stone crashed through the branches and the monkeys chattered amongst themselves, scurrying up the trunks and out onto the uppermost limbs. They stared down at me with their large inquisitive eyes, muttering to themselves. I aimed another stone and flung it, instantly renewing the chattering. Behind me I heard a low whistle. Erin stood at the edge of the porch, thermos in one hand, and a bag of sandwiches in the other. She smiled.

It was time to go.

We drove through town, past the glitzy department stores, past the old railroad station, and beyond towards the outskirts of the city. The streets were still deserted with only a few early risers to be seen. This was my town, Pietermaritzburg. I had grown up here, and the streets and buildings felt as familiar to me as an old friend. Before long the buildings thinned out, and we were into the country with rolling hills gently undulating as far as the eye could see. It was winter, and an occasional pocket of frost stood stark against the brown landscape. The early light was filtered by low cloud, and in some places a gentle mist hung in the valleys. It dampened the windshield and made the road slick, but the sun would soon be up to burn off the remaining wisps of mist and pockets of frost. A hot day lay ahead. The truck was running well, and I felt a warm contentment in anticipation of another adventure.

Africa is ingrained in my spirit. I can feel the rhythm of the land resonate through my blood the moment I return to the country. Each continent has its own beat, and you can feel it as soon as you arrive at the airport, but only the land of your birth has that special effect on your being. All those formative years seep slowly into your system and are instantly triggered the moment you breathe the air that you were born into.

I felt the rhythm resonate with each mile.

It had been five years since I had left, and I was returning to South Africa to show my first wife the land that I loved so much. I wanted her experience to be pure Africa — the people, the countryside, and the downside, and there was no better place to find these things than the "wild coast." There are few places left on the planet that remain completely untouched by man's indiscriminate need for space. The "wild coast" is one such place. It runs two hundred miles from the Natal border south towards the Cape Province, and forms the coastline of the Transkei, one of South Africa's so-called "independent homelands." It remains unspoiled for one simple reason: it is largely inaccessible. All the roads that run to the coast, run perpendicular to it, and it is a long, dry, dusty slog to get there. Only true believers make the effort.

The sun broke the horizon behind us, and before long it had climbed high enough to beat down on the car, warming the inside. I rolled my window down, and the cool, sweet-smelling air drifted in as we made our way through the rural villages of western Natal. The mud huts were painted in bright colors giving them a gaiety that belied the surrounding poverty. Barefoot kids scrambled down to the roadside to watch us pass. They stood and waved, some of them cupping their hands in hope of a handout, but we blew by, intent upon getting to the Transkei border before lunch. I could see the look of dismay on Erin's face as we passed without stopping, but after a few more miles and hundreds more kids she knew, as did I, that helping out a single family would hardly dent the overwhelming poverty. We swallowed our guilt and kept on going.

The road was a narrow, single-lane highway, and occasionally a truck carrying livestock would pass us heading towards town. The trucks were piled high with their precious cargo balanced precariously. The vehicles swayed dangerously. I pulled over onto the dirt shoulder to let them go by, and felt the wind buffet our car as they passed. The smell of scared animals hung in the air long after they were over the horizon. Occasionally we would encounter a piccanin, a young farm boy, herding a flock of sheep down the road, and we would slow down to let them pass. Mostly, though, it was just a rolling road heading west.

By late morning we made the border of the Transkei and pulled up to the disheveled customs buildings. A uniformed guard strolled up to the car picking his nose and sweating profusely. Customs were a mere formality and passage was automatically granted to all South African citizens. We did not expect any problems. Erin was traveling on an American passport and I had been using my British passport for the past few years. Neither one would present a problem.

"E' passiporti please." The guard extended a sweaty hand. I reached into my bag for the passports and a travel permit, and handed them over. We sat in silence while he studied the documents. I noticed his pistol bulging in its holster. The guard squinted at each stamp, fingering the paper and flipping the pages until he came to the photographs.

"Where you going Sah?" he asked. I explained that we were headed to the "wild coast" for a few days.

"And whose *bakkie* is dis?" he asked, using the Afrikaans word for pickup truck. I told him that it belonged to my brother, and watched while he wandered around the car inspecting the gear in the back.

"You want to sell it?" he asked. "I give you three thousand rands today."

"No thanks," I replied, "I can't do that. It's not my car." He stopped at Erin's window and peered inside.

"You sell dis *bakkie* to me today," he insisted, wiping his hand across his nose.

"I can't," I said. "It's not mine to sell. Besides we need it to get to the coast." He sniffed and spat on the ground.

"Very well then. You sell dis' *bakkie* to me now otherwise I will not let you pass," he continued, his voice rising with an edge that made me uneasy.

"I can't sell it," I insisted. "It's not my car. I don't have the papers." Arguing was pointless, judging by the guard's demeanor. I sat in silence and watched him. He wandered around to the rear of the truck lifting the tarp once more to inspect our gear, and then returned to the customs building. "What now," I thought? The heat was making it uncomfortable in the car, and I was getting anxious and agitated just sitting and waiting. "Christ, this is ridiculous," I muttered under my breath. I was keen to keep moving.

After twenty minutes he reappeared.

"You Missus," He gestured to Erin. "You get out of the car." She looked at me and I nodded. "You stan' against the car and wait, and you Mister, you stay in the car." I sat quietly behind the wheel feeling the level of agitation rising.

"Where you from Missus?"

"America," she replied.

"You like men?" he asked. Erin did not reply. "You like American blek men?" he continued. Erin said nothing. "You tell your boyfriend to sell me this car." What was it about the car I wondered. Why was he being so insistent? Three thousand rands was about half its value, but that was probably not the reason. He was just out to harass us and exercise his newly granted powers. The "independent homelands" were a lousy idea anyway, I thought. Give a guy a gun and a uniform, and he loses his sense of reason.

"You like blek men?" he asked again moving closer. Erin stepped back.

"Yes," she said softly. "I like men." The guard beamed.

"Now we are getting somewhere," he said. I glanced in the rear-view mirror and noticed another car approaching. The guard saw it too and spat on the ground.

"Shit," he said. "You get back in the car Missus." He handed her the passports. Without further comment he raised the gate and waved us on. A lime-green VW beetle pulled to a stop at the gate, and I looked back in time to see the guard spit one more time, adjust his nuts and stroll up to the driver's side window.

We pulled away, and a few miles later the tar abruptly ended, and the road turned to dirt. The pickup drifted sideways as it hit the new surface, and dust billowed out behind us. It was a while before either of us spoke.

"What was that all about?" Erin asked. I shrugged and stared intently at the road. The guard had been acting strange, but strange guards were nothing new in Africa.

Who was it that once said —"*power corrupts, and absolute power corrupts absolutely*?" I wondered. The "independent homelands" were a bad idea, bound for certain disaster. The guard was just a small example of what might happen. I was looking forward to the day when there would be an equal vote in the country. It had to happen, and the sooner it did, the better it would be for everyone.

"I don't know," I said. "Too much power and not enough brains." I watched the dry land pass by and wondered about the future. I wondered what would become of South Africa? What would happen to this beautiful land? What would happen to my country? I had once been told that rape was the national sport of the idle African, but didn't subscribe to that kind of thinking. Still, the guard had unnerved me. I was pessimistic about the future.

"What do you think would have happened if the other car had not arrived?" Erin asked. I stared at the road and didn't answer. I had no idea. The incident made me feel uncomfortable, but I was not going to let it ruin my day.

We stopped in Umtata for lunch. Umtata is the capital of the Transkei and houses the puppet government. We picked up a six-pack of Lion Lager, and some biltong to go with the sandwiches. South Africa is famous for its dried meat, and a slab of ostrich biltong would be a great addition to the lunch that Erin had packed. We were eager to make Magwa Falls by dark and piled back into the truck, which by now was a dull red, thickly coated with Transkei dust. I could feel it caking the back of my throat. Erin cracked the top of a Lion Lager and passed it to me. I drank gratefully, and the beer washed away the film of dust that coated my tongue and the bad taste left by the guard. I steered the car with one hand, clutching a thick sandwich and the beer with the other, while the truck drifted around the corners in a billowing of dust. "Bakkie surfing," it's called, and to an African kid it's about as close to heaven as it gets.

The afternoon drifted lazily by. The dirt road gradually narrowed as we got further away from town, until it was a track barely wide enough for a single car. There was no other traffic on the road, so it didn't matter. The countryside was flat and featureless and dotted with small clusters of huts. There was no organized grouping to the villages. They had sprung up in a haphazard way, cluttered around what water they could find. Scrawny farm animals scrounged whatever nutrition they could from tufts of dry grass. I was amazed that they survived in such a harsh place, but Africa is all about survival.

We finished the six-pack and stopped for another at a roadside stand. The shelves were mostly bare, and everything was covered with a fine film of dust. In the rear of the store a refrigerator chugged desperately, struggling to keep its contents cool against the heat of the day. A radio played, and the attendant sat bored by the door. I asked if there was a place to buy petrol. She waved her hand in the general direction of the coast and shrugged.

"Somewhere down there Baas you can buy petrol." She didn't seem too certain. I wished I had brought along a spare jerry-can, but it was too late now. I paid for the beer and headed back to the car. The air had turned still and hung heavy. Towards the west I noticed a bank of cumulus clouds rolling in, their anvil shaped tops reaching up beyond the cirrus.

"I think we might be in for some rain," I said to Erin. I could feel the moisture wrap itself around everything, and the air was thick with dust and humidity. Erin dozed off from the heat and beer, and the clouds closed in as we pulled into Magwa Falls.

The sign at the gate might once have greeted all visitors gaily, but now it hung lopsided on a wooden post with the paint peeling. We passed over the cattle grid, down a tree lined avenue, and up to the main reception area. The sign on the door was faded, but it too beckoned a welcome, and while Erin stayed in the truck to watch our gear, I went inside to register.

The reception area was cool and dimly lit. As my eyes adjusted to the low light, they focused on a rundown counter with a few faded promotional posters from a bygone era hanging on the wall. There was no one around. I rang the bell and waited, but no one came. I rang again, and then noticed a door off to the side. I knocked tentatively, and when there was no answer, I quietly pushed it open. The room was dark with all the curtains drawn. A faint wheezing and puffing emanated from the far corner, and I could just make out the slouched form of someone in a chair. The receptionist was there fast asleep, oblivious to my presence. Her huge breasts rose with each wheeze, and fell gently with each puff.

"Hello," I called, "hello, hello." The breasts rose and fell without interruption. "Excuse me ma'am. Hello, hello." No movement. The wheezing continued while her ample mounds of flesh rolled in rhythm. I looked around the room, and then went back out to the reception area and banged the bell. Through the window I could see Erin waiting patiently. I decided that I had better wake the receptionist, and tentatively approached the wheezing body, gently nudging her shoulder and calling out a soft "hello," but the body only wobbled to my touch. There was no response. I prodded again, and my finger sunk into the fleshy folds. All of a sudden her eyes flashed open and she jumped to her feet.

"Sheeeeet Baas, wad you doin'?" She stared at me with wide eyes and then slumped back into the chair. "Sheet Baas you dam' scared the sheet out of me. What you want here?"

"I'm here to check into a room," I stammered. "I have a reservation."

"Sheet Baas," she said again, using a servants term for an employer. "You don' need a reservation around here. No one comes here any more. If you want a room you jus' take a room." She rose slowly from the chair, a lot slower than the first time. "You wan' a room Baas you can have any room. Twenty rands for the night." I followed her out to the reception area and pulled a twenty-rand bill from my jeans pocket. She shuffled over to the counter and reached for a bunch of keys.

"What do you mean no one comes here any more?" I asked. I was looking for a quiet place, and it seemed as if I had found one. Thunder rumbled in the distance and the windows rattled.

"Dis place is under new management," she said. "Since the Transkei got independent, dis place been under Government management. You take any room," she said, "any place you fancy. Dis key fit them all." I thanked her and took the key. The air outside was cool and fresh after the stale smell of the reception.

Erin smiled as I approached the car.

"Where have you been?" she asked, "I was worried about you." I looked at her happy face and thought I noticed a look of anxiety cross her eyes. Perhaps it was just my imagination. "Is everything OK?" she asked.

"Oh, fine, yea," I replied. "We have a room for the night." For the first time I noticed an eerie, still feeling in the air. Not like the calm that precedes a storm, but an ominous, languid feeling. I shivered despite the warm temperature. Something did not feel right about the place. The air was too still. There were no birds. No sounds at all. Nothing was moving. In the distance I could hear the surf crashing onto the beach, but it was only a muffled roar. We drove down the dirt road towards the huts where guests stayed. There was no one around. The road reached the edge of the beach and the surf thundered louder. The sand was a brilliant white, and the warm blue ocean crashed onto the shore in a frothy wave washing up to the high water mark. Beyond the breakers I saw white caps on the ocean and felt another unease as I realized that not fifty yards offshore a strong wind was blowing, but where we stood it was calm. Flat calm. Erin stared at the whitecaps, and then looked at me.

"Weird," she said.

"Very strange," I agreed. We pulled up to the first hut, a rondavel, as it's called in Africa. It was round with white-washed walls and a thatched roof, and was close to the beach. The wonderful sound of the ocean would lull us to sleep. I immediately felt better as that small part of being African connected with the remote setting and the rondavel. It had been a long day and I was looking forward to a swim and good dinner. It felt good to be back home with the wild coast-line bursting with energy. Maybe a good swim and a decent meal was all we needed.

Thunder rolled overhead and rumbled a low bellow that echoed across the dry land.

"We'd better be quick if we want to swim before the rain comes," Erin said. We piled out of the truck, and I opened the door to the rondavel. It was cool inside, but smelt damp and stale. The light did not work, and Erin pulled the curtain aside to let the outside light stream in. The place was a mess, with the bed unmade and clothing scattered on the floor.

"It looks as if someone is staying here," Erin said. "I thought you said that we were the only guests."

"That's what she told me at the reception anyway," I replied. "Let's try another hut." We walked over to the next hut and went in. It was bright inside, but smelt worse than the first one.

"Maybe it just needs a little air," Erin offered. I went into the kitchen looking for the source of the smell and found a pot bubbling on the stove. The stink in the tiny room was overwhelming.

"This is really strange," she said. "It looks as if there is someone staying here as well, and whatever they are cooking smells disgusting." Her American nose was sending strong signals to her brain. There was no way she would stay in this hut. The pot bubbled away, and as I lifted the lid to peek inside, a cloud of foul smelling steam escaped. The steam dissipated and I looked into the pot. Staring back at me was the half-boiled head of something. The front teeth were ex-posed. They were extra large and white and incongruous against the darkened eye sockets. I slammed the lid down as the hairs on the back of my neck stood strait out.

"Shit, let's get out of here!" I bolted for the door. From where Erin was standing, she had not been able to see into the pot, but she knew that it wasn't good. We ran back to the car.

"Let's get the hell out of here," I said. "This place gives me the creeps. There may still be time to find somewhere else before it gets dark." Erin didn't need encouraging. We raced back up the dirt road not bothering to stop at the reception to pick up my twenty rands, crossed over the cattle grid, and drove back to the main road. My heart was pounding, and the hairs on my neck were still standing on end. There was no way we were going to stay one moment longer. I had an eerie feeling about the place from the moment we drove in. Our problem was going to be finding another place to stay before nightfall. Accommodation was scarce, and the sun was already waning towards the western horizon with rain imminent. I pulled the truck to the side of the road and grabbed the map. Umzimvubu was down the coast a little, not far as the crow flies, but there were no roads running along the coast. We would have to drive the better part of an hour inland before finding a crossroad to link up with the road running to Umzimvubu. It was our only choice. Erin opened a couple of beers, and we headed back the way we had just come.

"That place was downright spooky," she said. I agreed and gunned the engine.

We would have to hurry.

Sunsets in Africa are like no other on earth. The air is filled with dust and moisture, and the sun's rays reflect off the minute particles. The landscape is brown and dry and absorbs the late afternoon sun with warmth and beauty. We were being treated to one of the best, enhanced by an approaching thunderstorm. The clouds were black, laden with rain. We drove in silence watching the fingers of light paint the sky. It would be dark by the time we arrived in Umzimvubu, but I was more concerned about beating the fading light to the crossroad. There are few signposts in the Transkei, and finding the road would be difficult even in daylight. We pressed on steadily as darkness approached.

I glanced at the gas gauge and cursed under my breath. The needle hovered just above red. We should make Umzimvubu without any problem, and I hoped that someone there would have a few gallons to sell us. I didn't mention anything to Erin. She had been a good sport all day, but I could sense her unease. I was feeling a little uneasy myself, and I was in a familiar environment.

"Beautiful sky," I muttered. Just then, large splats of rain fell heavily onto the hood of the car. They dissipated the dust on the windshield, immediately forming small rivers of mud. The warm drops drummed the roof, and in a few moments the din inside was too loud for talking. The wipers ran at full speed. It was hardly possible to see more than a few feet in front of us, and I prayed silently that there would not be any hail. A late afternoon hail storm can be a spectacular sight, but devastating to anything in its path. Thunder rolled across the heavens, and the dry earth drank up the moisture.

We plodded on looking for the crossroad.

As suddenly as it had started, the rain quit. Storm clouds lined the horizon and I knew that the worst was still to come. Up ahead I saw a small road running perpendicular to ours. It had to be the crossroad. I stopped the car, checked the map, and then pulled onto the intersecting road.

Road was probably a misnomer. In Africa the roads sometimes become paths, and other times they are just gullies that your tires fit into. This one started as a road, but after a few minutes began to fade. The dirt turned to a path, and soon the path turned into two gullies. Two narrow mud-filled gullies. I thought about turning back, but the level of the gas-gauge convinced me to press on. If we turned back we would not make Umzimvubu that night. Fortunately the gullies gave way to a wider track, and after a while it resembled a road again. And then the rain resumed. It came down with less urgency than the first foray, but it was still a heavy downpour. At times we were driving off the path when I would notice tracks running alongside of us, and pull back onto them. Night had fallen, and the headlights reflected brightly on the drops of rain. Every now and then a rabbit would stand frozen in the beams, and at the last moment it would regain its senses and bolt off into the black. Erin clutched the door handle, holding on as the car slipped and slid in the mud. My hands were glued to the wheel, palms sweating. I felt a slight chill despite the tropical air.

Our situation was deteriorating rapidly, and our options were getting more limited. The gas needle hovered just above empty, at times dipping below. The crossroad was a little over ten miles long, or at least that was how it appeared on the map. I was starting to feel that we were on the wrong road. We had been driving for almost an hour, and there was still no sign of the road leading to Umzimvubu. We slipped and slid and sloshed and bumped in the pitch black finding no trace of civilization, and no other road. The wind picked up and buffeted the car while the rain pelted loudly on the hood.

I was clearly worried, and Erin to her everlasting credit, was biting her tongue. I knew that she was scared and I felt scared myself. We were well beyond the point of turning back. The level of gas in the tank would barely get us off the crossroad, and definitely not all the way back to Magwa Falls.

We pressed on.

The rain continued to fall heavily. Without a moon, it was the darkest night I could remember, and the Transkei is so flat and featureless that there were no landmarks by which we could gauge our progress. Suddenly I noticed a break ahead, and with a bump and a skid we landed squarely on a larger crossroad. It ran perpendicular to ours — in the direction of Umzimvubu. I stopped and checked the map. This time I was sure that we were on the right road and swung the truck, pointing it south once more. South in the direction of the coast. Umzimvubu looked to be about twenty-five miles away, and with a little luck we would be there by 10 p.m. It had been a long day and we were ready to get out of the truck and into a warm bed.

The new road was a lot better than the cross-track, and other than poor visibility, we were making good progress. The truck was sliding in the mud, and I was having a hard time keeping it on the road. At each bend the rear-end would slide out, and one time it swung all the way around until we faced back the way we had come. It might have been a problem had there been other traffic, but we had not seen another vehicle since leaving Magwa Falls. We passed small clusters of huts and could smell the wood smoke from their cooking fires. I envied them, and thought of families gathered out of the rain, cooking big pots of food over open fires. I was hungry and longed for some place warm and dry.

"You OK?" I asked Erin. She nodded slightly.

"Better now than I was on the other road," she replied. "I was scared back there. I didn't think we were going to make it." Her face was pale and drawn. "Were you scared?" she asked. I nodded. It was obvious that we had been in trouble.

"It's OK, Erin," I said, "we will soon be in a warm place, and out of the rain."

"I don't care if Umzimvubu is like the other place," she said flatly. "We are not leaving."

"OK. Me too." I had no intention of doing any more driving. If we made it, it would be on fumes only. The rain continued pelting down, the drops reflecting brightly in the headlights. The visibility was almost zero when the road suddenly took a sharp turn to the right. Out of the corner of my eye I saw the track heading up a hill and reflexively swung the wheel, slamming the brakes. The car skidded sideways, and then spun a full three-hundred sixty degrees. I shut my eyes and hung onto the wheel as we took a violent bump and then became airborne, before making a soft landing in a ditch alongside the road. The engine screamed and the wheels spun violently, kicking up mud. My foot was jammed on the accelerator. I pulled it off and the car stalled. Suddenly it was very quiet. I looked over at Erin. Her face was white and she looked vulnerable and scared hanging onto the door handle. All of a sudden tears popped from both eyes. They ran down her cheeks and dropped onto her lap. She sat there looking at me and I hung my head.

We were now in real trouble.

"Shit," I muttered. "Damn, stinking, damn. How bloody stupid." I cursed at the rain and at the mud. "What the hell are we going to do now?" I sat with my head resting on the wheel and stared into the dark night. The headlights were pointing up at the sky fading out into the inky black. Beside me I could hear Erin crying softly.

"What a bloody idiot I am," I said. "What a stupid fool." It had already been a tough night and we were now in a real jam. How were we going to extricate ourselves from this situation? I turned off the headlights and we were pitched into complete darkness.

"Turn the lights back on," Erin said. "I'm scared, it's too dark." I flicked them on again and the beams pointed skyward. The rain had stopped momentarily and I rolled the window down. I could smell wood smoke in the air and guessed that there must be a settlement nearby. Perhaps I could walk and get help, but I knew that Erin would not stay alone in the car and I wasn't too sure how safe we were anyway. We were most likely the only white people for fifty miles, and strange things have happened to lost travelers.

"We'd better save the battery," I said, and turned the lights off again. This time we were ready for the dark and it did not take us by surprise. I took Erin's hand in mine; it felt cool in my clammy hand. She squeezed it gently and in the darkness I could see a trace of a smile cross her face.

"This ain't no America," she said. "Where is triple-A when you need them?" I smiled back at her and wondered what to do next.

"There's not much we can do until morning. We're stuck in this ditch, and we'll need a tow to haul us out. I am sure we are safe here though." I stated the obvious, and added the bit about being safe more for Erin's benefit than mine. I had no idea if we were safe, but saying it out loud seemed to reassure us both.

"Feel like a beer and a something to eat?" Erin asked. "They are not that cold anymore, but it's better than nothing." She passed me a lukewarm Lion Lager, and cut a few slices of biltong. We sat quietly sharing the can and chewing on the salty meat. The night was quiet and the clouds seemed to be breaking up. Occasionally a weak half-moon would appear through a break in the clouds, and shed its watery light onto the landscape. The mud glistened and reflected the silver beams. On any other occasion it would have been beautiful to sit and watch the huge storm clouds roll by a waning moon, but this was not a night for cloud watching. We finished the beer and decided to save the last one for breakfast.

"Why don't we try and get some sleep?" I said. "We'll see what tomorrow brings." At least it was warm and dry in the car. We tilted the seat back as far as it could go, and I shut my eyes trying to block out our predicament. I must have dozed for a moment because Erin was suddenly shaking me.

"Quick," she said. "There is a car coming." I looked towards where she pointed and could just make out headlights winding their way towards us. I kicked my shoes off and hopped out of the truck. My foot sunk into a few inches of soft mud and squelched with each step as I made my way up to the road. As the car approached the driver flicked his beams on high and caught me directly in the glare. I waved and it pulled up alongside me. I felt a sudden pang of anxiety. What if these people were not friendly? The car stopped and the driver wound down his window. A lone black man sat behind the wheel.

"Hey Baas, wad you doing out here?" His white teeth glistened, and I could see that he was suppressing a laugh.

"You're in trouble man. Wad you doing with your *bakkie* in the ditch? That's no place to park." He laughed at his own joke and opened the door. He was a big man and more laughter rumbled deep inside of him. He shook my hand still laughing to himself, and led me to the back of his car. He opened the trunk, reached in, and grabbed a short length of rope.

"You the second person I pull from a ditch tonight. Five miles back there," he said, pointing back the way he had just come, "I pull another *bakkie* from the ditch." He laughed again and patted me on the back. "Don't worry Baas, I have you out in a few minutes." From the car I heard Erin in a small voice;

"Are you all right, Hancock?" she said. "Is everything OK?"

"We're going to get pulled out," I shouted back.

I plodded back to the car with one end of the rope, while my new friend tied the other end to his tow hitch.

"Sit tight, he's going to pull us out."

"Should I get out of the truck?"

"No," I said, "just hold on and we'll be out of here in a moment." I tied the rope around the tow hitch and sloshed my way back up to the road. The rope went tight, the engine roared, but our truck didn't budge. He gunned the engine again – still nothing.

The black man pointed towards the car.

"You better get down there and push," he instructed. I skidded down the bank, made my way behind the truck and leaned my shoulder up against the front bumper.

"OK," I yelled, and heard the engine roar. There was a soft sucking sound as the truck moved, and then suddenly it was free. I shoved harder, and with a pop the car was pulled backwards out of the ditch leaving me flat on my face in the mud. I looked up at the road and the man was standing there laughing.

"Come on Baas. Wad you doing down there?" I could see white teeth glinting in the moonlight.

"Shit," I said. "I have had enough." I heard more laughter as I slipped my way back up to the road. He had untied the rope from both cars and extended a hand to help me up the last few feet.

"Hey Baas where you goin' tonight?" I told him that we were headed for Umzimvubu for a few days and he laughed even louder.

"There's nothing at Umzimvubu any more," he said. "There's no place to stay there. Everything is closed."

I stared at him. "There's nothing in Umzimvubu? No rondavels?"

"No Baas, it's all closed. Since the new government took over, it has been closed."

"Oh man," I cursed softly "What now?" Our predicament had hardly improved since being pulled from the ditch. No gas for the truck and nowhere to stay. Erin had not heard the conversation and was sitting patiently waiting for me to come back to the car.

"Do you know any place to buy petrol? We need petrol for the truck." The man laughed again and pointed to the north.

"About fifteen miles up the road there is a pump. There is also a small hotel there. Not very fancy, but cheap. I must go now." He shook my muddy hand, jumped back into his car and sped off. I could hear laughter as he pulled away. In a moment his tail lights had faded and it was dark again. I climbed in alongside Erin and she grinned at me.

"What's so funny?" The mud was starting to cake, and dry bits flaked off in the car. Erin laughed and took my hand.

"So what did he say about Umzimvubu?" I told her about the hotel and we headed back the way we had just driven. The town would be north of the crossroad, and if our luck held we might just have enough gas to get us there. I drove slowly. Without the rain it was easier to see where we were going.

Half an hour later we pulled into a small village with no name and drove slowly down the main street. The town was deserted and very run down. A few ramshackle houses lined the only street, and half way down a sign read OTEL. The H was gone, but that was the least of its problems. The roof was sagging and all the paint had peeled off. I stopped out front and jumped out of the car.

"I'll go and get us a room," I said, and looked for the reception area. There did not appear to be one so I peered through one of the open windows. In the corner of the room a narrow cot was covered with a filthy spread. The room was tiny, and on a wooden bed stand there were a few candles and a box of matches. An unwashed smell permeated the place. We were both tired, but I could not see Erin agreeing to stay in that room. I hoped the good rooms were in the back.

I told Erin what I had seen and she still insisted that we should stop for the night.

"How bad can it be, Hancock?" she said. Maybe she was right. We found the reception area and let ourselves in. There was no one around. I banged on the window and called out, but no one came.

"This place is not so bad," Erin said. "I'm so tired that I would not even notice if the bed was hard."

"Hello," I called out. "Hello. Anyone here?" In the corner I heard a small rustling sound and glanced down. Two rats were chewing on an old newspaper, and then one of them scurried across the floor and ran across Erin's foot. She screamed and fled back to the car. I found her sitting with her door locked and her legs pulled up under her.

"Wewwe.. are not staying there," she stammered. "Let's get the hell out of this place. I want to go home," she wailed. I agreed. It was time to bag the vacation and get back to civilization. We might even make it home before dawn.

"Let's go," I said, and then remembered the gas. We had been running on fumes for a long while. We were not going anywhere until we found a pump.

We drove up and down the main street looking for something that might resemble a petrol station, but without luck. On the edge of town I saw some kids hanging out and stopped to ask them if there was a place to buy petrol. They directed me behind a rundown warehouse where an old fashioned gas pump stood looking like it had not been used in years. The kids told me that old Mr. Mbeki was in charge of the pump, and he lived a short walk down the dirt track that ran adjacent to the warehouse. I looked where they pointed and the track faded into the dark night. The storm clouds still hung low and the sliver of moon barely shed light. I told Erin to wait in the car and lock the doors. We had to get gas and if old Mr. Mbeki had the keys, then I would have to find him. I fumbled through our bags, found a flashlight and set off down the path. It was slippery from the rain, and narrow. Some branches grew across the path and I had to duck to get under them. I was glad of the flashlight, but even its faint beam did little to calm my nerves. The night was so black that the feeble beam faded out a few feet in front of me, swallowed up by the darkness. I began to wonder if the kids were sending me on a wild goose chase when my ears caught the sound of muffled laughter.

It had to be Mbeki's place.

Back in the car Erin sat quietly with her feet tucked up under her, all doors locked. She left the engine running and every few seconds glanced nervously down the path hoping that I would reappear. It had been a long day and her imagination was working on overdrive. The skewed American view of Africa being a wild and savage place was haunting her, but so far without reason. With the exception of the guard at the border crossing, all the people that we had encountered in the Transkei had been good to us. Still, the shadows bothered her, just as they were bothering me.

The muffled laughter grew louder as I stumbled my way towards the sound. In a clearing ahead I discovered a small cluster of huts grouped together against a low cliff. There was a pungent smell of wood smoke in the air. I guessed that dinner would be well over and the fires were being kept alive to keep out the dampness. It had started to drizzle again and the drips ran down my face. My wet hair hung in my eyes. I wasn't sure what to do. I was nervous about knocking on the door and asking Mr. Mbeki to crank up his pump, but there was no way that we were going to return to the hotel, and the car for the night was not a great option. I wanted to go home.

The loudest voices seemed to be coming from the main hut and I cautiously approached the front door. I listened for a moment and then called out;

"Hello. Hello. Mr. Mbeki?" Immediately the voices inside hushed. "Is Mr. Mbeki in there please?" The door swung open and a teenage boy stood in the entrance naked to the waist. He stared at me. Behind him the room was filled with smoke, and I could make out a number of people sitting on a bed.

"Is Mr. Mbeki in?" I stammered. The boy moved aside to let me in. An old man was sitting in the middle of the bed with a number of small children snuggled around him. He was wizened with a scraggly gray beard. Most of his front teeth were either yellowed or missing.

"Yebo Baas," he said, "I am Mbeki." The children sat silently staring at me with big brown eyes and snotty noses, and cuddled closer to the old man when he spoke. It was very hot and very smoky in the hut, and my eyes smarted. I explained that I needed to get some petrol and Mr. Mbeki nodded and asked me to sit down. I looked around for a chair, but he patted the mattress motioning for me to join him on the bed. There was long pause while I sat uncomfortably waiting for his answer. I was worried about Erin alone in the car and wanted to get moving, but Mr. Mbeki had the key to the pump and I was not about to rush him. In broken English he asked why I was there and why I needed to get petrol now and not in the morning. I lied and told him that my wife was feeling sick and I wanted to get to Umtata. He nodded thoughtfully and said something to the teenage boy who disappeared into the next hut. Mr. Mbeki rubbed his chin, deep in thought, and then rose slowly from the bed. His small frame was stooped and he walked with an awkward shuffle. The teenager returned with a bunch of keys and Mbeki nodded to me.

"Follow me," he said, and we set off back up the hill. I

followed his frail frame attempting to light the path in front of him with my flashlight, but Mbeki knew his way. He had obviously walked the path many times. We inched our way along as he chose his footsteps carefully, and after a while the smell of the huts disappeared and we approached the top of the hill. I could see the pump and hear the truck running.

Erin jumped as I approached the car and smiled with relief when I told her that we would be able to get gas and keep moving. Pietermaritzburg was still a long way away, but now we had a chance of making it back home. I swung the truck around and shone the headlights at the pump while Mbeki fumbled with the lock. And then the engine spluttered and died. It was suddenly very quiet – we were out of gas.

Mbeki fired up a small generator. The noise broke the stillness of the night and the pump spluttered. He smiled thoughtfully while pumping, and beamed gratefully when I paid him three times the going rate.

It was 11:30 p.m. and still drizzling.

I studied the map and found a road that would take us back to Umtata. The road was slippery and driving was difficult with poor visibility. Thunder rumbled in the distance and the smacking of the wipers lulled us as we drove on through the African night. Erin soon fell asleep while I fought desperately to keep my eyes open. I was tempted to pull over and rest, but opted instead to press on.

A little after midnight, it started to rain again while sheet lightning lit the sky. The road had turned into a narrow wet gully. I checked the map. We seemed to be on the right track. I thought I could make out a light in front of us and rubbed my eyes, sure they were playing tricks on me. We had not seen another car since being pulled from the ditch, but as we got closer I could make out the faint tail-lights of another vehicle. I was happy to have someone to follow because the visibility was all but gone, and the road had started to drop off down a steep hill. I was worried about skidding off the track. I slowed to a crawl and followed the two red lights as we slipped and slid in the dark. I was glad that Erin was asleep. At times I lost sight of the car ahead only to find it again around the next bend.

Suddenly the road leveled out and the car ahead stopped. I stomped on the breaks and skidded to a stop inches from the car in front. The headlights reflected on rapids directly in front of us. The river had broken its banks and was flowing strongly. The driver hopped out and walked to the waters edge to inspect the flow. He waved at me, and then, without hesitating, jumped back in his car, gunned the engine and sped into the water. Waves splashed and the car drifted sideways. In a moment he was on the other side. He leaned out of the window and beckoned to me to follow. I glanced over at Erin who was still asleep, gulped and gunned the engine. The truck hit the rapids on the upstream side of the bridge, and Erin woke just as the first wave crashed over the car. The wheels lost all traction and suddenly we were afloat.

"Hancock, what the hell are you doing," she yelled.

"Hold on," I said. I clutched the steering wheel and prayed. The truck bobbed in the rapids, but there was enough momentum to carry us to the other bank. The tires spun until they caught, and we skidded up the other side and pulled to a stop alongside the other car. My feet were in a few inches of water so I opened the door to let the flood pour out. Erin glared at me – the rain kept pounding down.

In the car beside us were five well-dressed black men

returning from a party somewhere. They were all smashed, and the three in the back were fast asleep. The smell of cheap whiskey hung over their supine bodies. The driver was dressed in his Sunday best with mud caked to his knees. His once smart shoes were now blobs of clay, and his jacket was wet and covered with mud. He was grinning heartily, flashing a perfect set of teeth, and enjoying every minute of this great adventure.

"That was the easy part Baas," he told me. "The river is no problem except once my brother was washed away." He waved in the general direction of downstream.

"The hard part is dis' hill." He was right. On the way down to the river the car had been sliding all over the place, and it looked as if the hill ahead was even steeper. I shook my head and wondered again what else could possibly go wrong.

"Don't worry Baas, you jus' need speed. Lots of speed. And a good driver." He grinned and then yelled at the three sleepers in the back. "And no passengers." The three dazed passengers tumbled from the car and staggered around in the rain trying to gain their bearings. It was obvious that they had been through this routine before because they set off up the hill leaving only the driver at the bottom with the car.

"You Baas," he said, pointing at me. "You follow them, and you Missus," he said, pointing at Erin. "You stay in da car." He seemed sure of himself and I took off up the hill. It was very steep and the rain had washed away a portion of the road. We rounded a bend and lost sight of the two cars. My companions looked miserable as their wet suits hung loosely and mud caked their legs up to just below their knees. We stood in the rain waiting patiently and I wondered what the plan was. In the distance I heard an engine revving. A moment later headlights appeared at the corner, and then the car hurtled into view at full speed. It slid sideways at the bend and almost ditched itself, but the driver was obviously practiced and deftly steered it back on track. The engine raced and the wheels spun wildly, flinging mud everywhere. The car pulled abreast of us just as its momentum was waning, and the four drunks and myself jumped in behind and started pushing. Mud squirted out from each tire and we slipped and slid and pushed and shoved the car to the top of the hill. With a roar, it crested and took off on the flat. The driver pulled over to the side and then walked back to join us.

"Good job," he said. "Now your car Baas. Jus' be sure you have a lots of speed." I walked down the hill to the river and found Erin sipping on the last Lion Lager. She shook her head at me.

"You are a mess, Hancock." I looked down at myself. Mud was plastered everywhere. I tried to rinse some of it off in the river, but the flow was running swiftly and I was worried about being washed away. Now it was my turn for some deft driving. I climbed in beside Erin and revved the engine. We took off up the hill with as much speed as I could coax in the slippery conditions, and approached the first bend with a good head of steam. The car drifted sideways as we took the corner, and I gunned the engine. With the car still drifting the added power thrust us into a spin. In a second we were pointing back down the hill. Erin yelled, and I hung on as we skidded towards the edge of the road. The car stopped in time, but we had lost our momentum. We slid back down the hill to the bottom. The second time I took off a little slower, and we hit the corner just right. The car drifted slightly and then straightened out as I accelerated for the stretch to the top, but we had lost our momentum. The wheels skidded and slipped and we ground to a halt. I could vaguely make out five shadowy figures huddled against the rain about a hundred yards up the hill.

"Shit," I cursed, "we're never going to make it up this damn hill." I looked at Erin in the dark and could see lines of fatigue around her eyes. She started to giggle and then laughed out loud. What a night. What a dumb, stinking, strange day it had been, and the end was still not in sight. In a few hours the sun would be up and perhaps then it would be dry enough to make it up the hill. Erin's laughter was infectious and soon we were both laughing out loud. Tears ran down my face, but they were tears of frustration more than joy. I glanced over Erin's shoulder to see a black face peering in. When he caught my eye his face broadened into a huge grin and he shook his head at us.

"I drive Baas," he declared and opened my door to get in. Without much thought I hopped out while he took off down the hill with Erin still in the passenger's seat. The lights turned the corner and we were plunged back into darkness. I shivered from the cold and waited for the lights to reappear. I heard the engine roar and then saw the truck hit the bend with a full head of steam. They drifted sideways and then straightened out. The truck passed me at full speed and to this day I have a clear picture of the driver hunched over the wheel with a maniacal look in his eyes, and Erin alongside him, sitting bolt upright, staring straight ahead. Her pale gentle face contrasted with the determined look of the driver. In a few seconds they were gone and the noise of the engine died away. They made it all the way to the top without a push. I ran after them, and by the time I got to the crest the other car was gone. I opened the door and jumped in. Erin glared at me.

"Hancock," she said. "You now owe me one, and it is going to take you the rest of your life to pay it off. You're a bloody mess," she declared. "Just take a look at you." She had not been out of the car since we left Magwa Falls and was sitting clean and pretty alongside me. I had mud caked everywhere from the blobs that were once my shoes, to the crusted bits stuck in my hair. I covered my face with my hands and shook my head.

"What the heck," I muttered. "What the hell else can go wrong?" I pushed the car into gear and let the clutch slide. "Home," was all I said.

The road was flat and in reasonable condition and before long we passed through Umtata for the second time. The rain had tapered to a fine mist by the time the dirt road turned back into tar. The small store where we had bought the beer and biltong was dark, and the streets were deserted. Erin dozed while I struggled to stay awake until just before dawn when the border post came into view. The buildings were dark and I was tempted to drive right on through without stopping, but there was a single light burning in the guard post. Erin volunteered to take our passports in to have them checked. I was hardly recognizable and thought it wise to stay in the car. She took both passports and disappeared into the building. I tilted my seat back a little and dozed off. I dreamed of mud and boiling heads and swift flowing rivers, and jumped suddenly when Erin shook me.

"You need to come in," she said. "There seems to be some problem with your passport and they won't let us pass until they talk to you." I staggered out of the car littering the path with flakes of mud as I made my way into the building. The walls were cold and intimidating. We passed an open cell and walked to the office at the end. Behind a small desk sat a huge Afrikaner, his expansive gut hanging limply over his belt, and his shaved head reflecting the bare light bulb.

"*Meneer* Hancock," he addressed me sternly in Afrikaans. "It seems to me that there is an issue here with unpaid taxes." He spat out the word *taxes*, and eyed me over the rims of his glasses. I said nothing, too tired and numb to respond.

"*Meneer* Hancock," he continued. "Were you not born in the Republic of South Africa?" I nodded. "And Meneer Hancock, it seems to me that you have been traveling for quite a few years, have you not?" I nodded again.

"So *Meneer* Hancock, I must ask you this very important question. Where do you pay your taxes?" I stared at him blankly. What on earth was he getting at?

"It seems to me," he stated matter-of-factly, "that you have been traveling for a number of years, obviously earning money, and probably not paying taxes." He spat the word out again, as if it was a bad taste in his mouth, and waited for my reply. My head was spinning and I could not think of a good answer. I stared dumbly at him and noticed a fine sweat breaking out on his brow.

He licked his lips.

"I need an explanation, *Meneer* Hancock," he said. Erin said nothing. I said nothing. The room swayed and the fatigue seeped into my legs. "I am waiting, *Meneer*." I glared back at him and then in a clear voice, answered;

"I do not have to pay tax on the income I receive. The money I get comes from my family. You see," I said, "my father sends me money to stay away from home because I am an embarrassment to the family." Erin shot a quick glance my way and the customs man burst into laughter. His body shook and his gut wobbled.

"*Meneer* Hancock," he said. "My job is very boring as you can imagine. But you have made my evening." He laughed again and I could see Erin visibly relax. "That was funny," he said. "Very funny. Very funny. But, *Meneer*, I am going to have to lock you up for the night anyway." He paused, waiting to see what effect his words would have on us, but we were too tired to care.

"Why?" I asked.

"*Meneer*, look at you." He lowered his voice so a softer tone. "You are a bloody mess. You can hardly stand up because you are so tired. If I let you go now you will crash your *bakkie* before the next bend. I'll put you both in a cell and you can rest until the new guard comes in at seven-o-clock." There was no point in arguing with him. He was right. He rose from behind the desk and pulled a bunch of keys out of the drawer. "Follow me please." He led us back to the open cell and stood aside as we filed in.

"You are the only ones here tonight so you have the place to yourself." He pointed to the two small cots. "They're too small for two people to sleep in," he said with a chuckle. "I know because I have tried." We both collapsed fully clothed onto the cots and in a second we were fast asleep.

It seemed to be only moments later when I heard the cell door rattle and forced an eye open. The customs man was standing there with two steaming cups of sweet tea, and announced that it was morning. He pointed us towards the showers and said that we would have to leave in fifteen minutes before the other guard came on duty.

"He's a sonofabitch," he declared. "He'd leave you locked up for a week, just for the hell of it." I was tempted to ask if it was the same black guard that we encountered the day before, but thought better of it. We showered, and feeling better stepped out into bright sunshine. All trace of the storm was gone and everything was washed clean. Someone had washed and waxed our truck and it shone brightly in the sunlight.

"What do you think?" I said to Erin. "Should we head back to the coast?" She glared at me without answering.

It was almost noon when we rolled into Pietermaritzburg and the city sweltered in the midday heat. We passed through town and then drove on north, up through the more affluent suburbs that surround the city. The air was cooler, and I could feel the temperature change as we arrived back home. The hydrangea bushes that lined our driveway were still in full bloom, and their sweet fragrance greeted us as I stopped the truck in front of the house.

It had been a long day.

SAME THING DIFFERENT

*"A man in jail has more room, better
food and commonly better company."*

Samuel Johnson

The evening light filters through my wet cabin window, and the reflected pattern plays a game of chase on my navigation station. I am amusing myself watching the water shadows dance back and forth. They have been at it for over an hour and I have sat and observed the game intently, even though I know what the outcome will be. The large squiggly shadow will never catch the small fat one, and before long they will both evaporate and become a distant memory. Still, it is a nice distraction from sailing the boat and I have managed to pass another hour at sea, all alone.

A quick glance at my GPS shows a speed over the ground of eleven knots. We, my boat and I that is, have become one and are eleven miles closer to Bermuda that we were an hour ago. A full hour further into this passage than we were when the squall passed through. An hour closer to the moment when someone else can do the thinking for me. But who's counting? Me, that's who.

Sunlight filters through transparent sails
— rare moments of pure magic.

The hardest part of this trip has not been the constantly changing weather and the vagaries of the Azores High. Neither has it been the solitude. I have enjoyed being alone. I just miss having someone else to think for me. I wish I could shut off for a while, just a short while, and let someone else make the decisions. Let someone else decide which way to go. Let someone else trim the sails and make the dinner and navigate and communicate. Let someone else keep an eye out for ships while I creep into my bunk and detach myself from the world for a while. The constant need to be alert has definitely been the hardest part. Since I left the Azores two weeks ago, it has been me, and me alone, making all the decisions, and it has been a twenty-four-hour-a-day task. There has not been one moment when I have been able to let my guard down and rest. You can't rest when you're sharing the ocean with ships and sailboats; most of which are not keeping a good lookout. When you ask a ship if they can see you on their radar, without exception they all ask me to wait a moment while they turn it on. It's not a comforting thought, nor does it allow for much rest on board.

Please don't think I am complaining though. I signed up for this adventure and I would not trade it for anything. Solo sailing has injected a necessary element into my sailing career and reinvigorated my interest in being offshore. It was becoming humdrum out here with other people. Once you have sailed around the world a few times, and crossed the North Atlantic more times than you care to remember, the thrill you once felt is lost, left bobbing in the wake of some long forgotten passage. Do the same Atlantic crossing alone, and suddenly it is a different story. I feel reenergized — more alive, more focused and more intent upon the work at hand. It's a new challenge, and there are a new set of circumstances to deal with.

It's the same thing, but different.

Do I miss the days when I could share the sunrise with my crew mates, and gulp down a bowl of freeze-dried food with my friends? Sure I do. Do I miss a cold wet sock in my face as the guy sleeping above me clambers to his bunk? Not a chance. Nor do I miss the tension that inevitably develops when you cram sixteen unwashed men in a small enclosed space, and send it plummeting down the face of a forty-foot swell. The only arguments taking place on this boat are between me and myself.

"Why did you ever think of doing this you fool," and "why did you head south of the rhumb line when it's (now) obvious that the north side was favored?" These disputes are always resolved in my favor, and we (my alter-ego and I), get along just fine most of the time.

No, I do not think I would trade this solitary existence for a boat load of problems. I will learn to deal with the decisions and live with the consequences. I will learn to pace myself so that I can make good choices whether it's two in the morning, or two in the afternoon. Time of day means nothing when you're alone. The days blend into nights, and the nights blend back into days. I am learning to live alone and love the life. It's a living worth loving – simple, focused and basic – with only myself to rely on, and only myself to blame. Solo sailing is the essence of a simplicity, and once you have experienced it for yourself, the addiction is hard to kick.

I know, I've tried.

SHADOWS (the postscript)

> *"All other creatures look down towards the earth,*
> *but man was given a face so that he might turn his*
> *eyes towards the stars and gaze upon the sky."*
>
> **Ovid**

We left our final anchorage in Newfoundland in the early morning. There was a low mist drifting across the bay and wisps of cloud hanging in the birch trees. A small boy rowed his skiff out to watch us leave. He reminded me of myself – a boy with dreams in his heart. As the anchor came up and we slowly motored away, I turned to wave goodbye, but the boy just stared after the boat. He too was traveling, his own thoughts taking him to far away places. Finally, almost as an afterthought, he waved back and seconds later was swallowed up by the mist.

We hoisted the sails and set a course for home. I was looking forward to seeing my family again. Traveling and searching my soul for answers is one thing, but being home is where my heart has always been. On the horizon ahead, icebergs stood like silent giants drifting south with the current. I whispered my dreams to the wind and quickly settled into the rhythm of life at sea. My journey was nearly complete – the one within and the sailing passage. I had visited my stories and rediscovered their lessons. I had learned from my mistakes and grown from each experience. I had learned from the many people that drifted into my life. Their memory remains like a shadow on my soul, and like shadows, they have come and gone. I have discovered that it's good to let them go. You can't cling to the past – it slips through your fingers like soft sand through an hourglass.

A chilly wind picked up and the new waves slap slapped on the hull. The rhythm resonated through the rigging and carried back to me at the helm. I watched the land sink slowly into the sea behind as we sailed away, and felt like the luckiest person alive. The horizon ahead was curved in a wide arc like outstretched arms. The future was fresh and new and full of possibility. It seemed like a good place to end this book, and so in doing so I leave you with one last thought.

Why not go out on a limb – isn't that where the fruit is?

This book is dedicated to my friend Ted Lonsdale – he too was a man of the sea.

"I really don't know why it is that all of us are so committed to the sea, except I think it's because in addition to the fact that the sea changes, and the light changes, and ship's change, it's because we all came from the sea. And it is an interesting biological fact that all of us have in our veins the exact same percentage of salt in our blood that exists in the ocean, and, therefore, we have salt in our blood, in our sweat, in our tears. We are tied to the ocean. And when we go back to sea – whether it is to sail or to watch it – we are going back from whence we came."

President John F. Kennedy
Newport, RI
America's Cup
Sept. 14, 1962

—————— **Acknowledgments** ——————

Many people have been kind enough to read these stories and offer constructive criticism, and I am grateful to them all for their input and comments.

I am mostly grateful for their encouragement.

The stories were meant to be personal narratives which I planned to one day pass along to my daughter to read when she was old enough. As I completed each story, I passed them around until there were people the world over reading and enjoying them, and waiting for the next one to come out. Their enthusiasm convinced me to combine them into a single book and publish it, so if you have enjoyed reading *Spindrift*, they are the ones to be thanked.

I am particularly grateful to Skip Novak for writing the foreword to this book. We have sailed many miles together and his competance as a sailor and writer have been a great inspiration.

There were three people in particular who read each draft of the manuscript and offered concise and sometimes scalding criticism. Kate Ford was the first to take a scalpel to it, Ann Harley cleaned up typo's and grammar, and Heather Ormerod help me put a shine on the final product. There were also a number of proofreaders, namely my father Basil Hancock, Steve Ellis, A.J. Caputo, Steve Polycyn, Larry Rosenfeld, David and Nonnie Thompson, Bruce Weaver Beth Leonard and J.T. Charles. Many thanks to you all for your time and effort.

I would especially like to thank David Weaver with whom I sailed the cold waters of Newfoundland. His help editing the final draft and his encouragement and insight wereinvaluable.

Finally, thanks to my wife, Sigrun ,who listened to each story as I read them aloud, and kept my ego in check.

Brian Hancock
Marblehead, Massachusetts
19th October, 1999

———— ORDER FORM ————

Telephone orders: Call toll free **888-639 2122**

Fax orders: 707-516 4946

By mail: Brian Hancock
Great Circle Press
9 Evans Road
Marblehead, MA 01945

Please send me _______ copies of SPINDRIFT, and inscribe each copy as noted below   (use a separate piece of paper if necessary.)

____

____

____

Name:____

Address:_______________________________________

City:_______________________ State: _________ Zip: ___________

Sales tax:
Please add 5% for books shipped within Massachusetts

Shipping:
Book rate: $2.00 for the first book and 75 cents for each additional book
(please allow two weeks for surface shipping)

Air mail: $3.00 per book

☐ **Check:**

☐ **Credit card:** Type: ______________________________

Card number: ______________________________

Name on card: ______________________________

———— **CALL TOLL FREE AND ORDER TODAY** ————

888-639 2122

or on-line at www.greatcircle.org

One of the most accomplished sailors in North America, Brian Hancock started his professional sailing career two decades ago and has logged over 200,000 miles offshore, including three Whitbread Round the World Campaigns.

In 1979 he raced aboard the winning yacht in the 13,000 mile Parmelia Race, sailing from England to Australia, and then in 1981 sailed as watch captain aboard *Alaska Eagle*, the first United States Whitbread entry. In 1985 he raced a second Whitbread, that time as watch captain aboard *Drum*, and in 1989 sailed the first part of the Whitbread that year as Sailing Master aboard *Fazisi*, the Soviet Union's first, and by happenstance, last Whitbread entry.

Brian is the owner of *Great Circle,* a light displacement Open 50 racing yacht built for shorthanded ocean racing. He has sailed across the Atlantic alone, and is an expert in single-handed offshore racing. He carved a niche for himself in the marine industry through not only his sailing ability, but through his ability to write about and articulate his adventures to audiences worldwide. He has spoken around the world to diverse groups ranging from yacht clubs and Maritime Museums, to inner city schools and business groups. He writes regularly for a number of well-known magazines, and can be found on the lecture circuit each winter sharing his experiences with audiences around the world. This is his first book. Brian is currently writing a novel.